TH

PHASE

CODE

THE PHASELOCK CODE

Through Time, Death, and Reality, the Metaphysical Adventures of the Man Who Fell Off Everest

ROGER HART

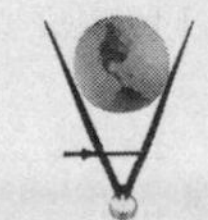

PARAVIEW POCKET BOOKS
New York London Toronto Sydney Singapore

PARAVIEW
191 Seventh Avenue, New York, NY 10011

POCKET BOOKS, a division of Simon & Schuster, Inc.
1230 Avenue of the Americas, New York, NY 10020

ISBN: 0-7434-7725-1

First Paraview Pocket Books trade paperback edition October 2003

10 9 8 7 6 5 4 3 2 1

Manufactured in the United States of America

For information regarding special discounts for bulk purchases,
please contact Simon & Schuster Special Sales at 1-800-456-6798
or business@simonandschuster.com

To the memories of
Robert Winfield Hart
and Woodrow Wilson Sayre

The mind, in short, works on the data it receives very much as a sculptor works on his block of stone. In a sense the statue stood there from eternity. But there were a thousand different ones beside it, and the sculptor alone is to thank for having extricated this one from the rest. . . . Other sculptors, other statues from the same stone! Other minds, other worlds from the same monotonous and inexpressive chaos! My world is but one in a million alike embedded, alike real to those who may abstract them. How different must be the worlds in the consciousness of ant, cuttle-fish, or crab!

—William James,
The Principles of Psychology, 1890

CONTENTS

CHAPTER ONE

THE ROUTE TRACKER

At last the meaning of these events is clear. I set out eager—yes, brash—in youth to explore polar ice caps, tangled jungles, and snow-clad mountains. I had visited every continent save one by the age of nineteen; I had a glacier in Antarctica named after me by the age of twenty. Then strange events in Tibet, Tierra del Fuego, Morocco, India, turned my course from explorer to scientist. I sought after the ultimate understanding of how our consciousness participates in the creation of reality. This is the adventure I share with you now—an adventure that changed my view of the world.

My story begins, when I was twenty-one, on an expedition to Mount Everest. Tibetan legend says Buddhist monks from the Rongbuk Monastery take the form of ravens and fly among the snowcapped Himalayas. If this is true, then they must have seen us on May 29, 1962, four black specks perched precariously in the glare of horrendous ice cliffs on the North Face.

We climbed without bottled oxygen, without porters, too exhausted to think straight. Woody Sayre, first on the rope, laboriously kicked footholds; Norman Hansen, twenty feet below, struggled under the weight of his pack; lower still, Hans Peter Duttle gasped for breath. I brought up the rear, bent over my ice ax, heaving long breaths, eight of them for each step.

I heard a sharp report on the North Peak. I twisted my body

to the right and focused on massive edifices of ice that had broken free, toppled, and now were careening down the slopes below the rocky peak. They shattered into pieces on a ledge, leapt free, and floated silently in thin air. The roar arrived after the vision of impact, like thunder after lightning. The chunks of ice splashed down, picked up speed, cascaded through a network of cracks, and trickled out onto white fans at the base of the rock face.

The hood of my parka reverberated steadily in the incessant wind. Slowly, one foot after the other, I kicked into the soft snow. Kick, step, then eight breaths leaning on my ice ax. Kick, step, eight breaths. The rhythm became a monotonous dirge. Norman and Hans Peter advanced even more slowly than I, and the rope in front hung limp over the sticky snow.

Above and to my left, clouds streaming north from Nepal slipped past the colossal prow of Everest's main peak. When I leaned back to study the clouds, they seemed to stop moving and the black hull of Everest plunged away from me, throwing me off balance. I threw up my arms, plunged my ice ax up to its hilt, and grabbed hold.

Don't look up, focus on the steps, I told myself. With only ice slopes and white clouds above us, it seemed as though the North Col, the ridge between the two peaks of Everest, was just over the next rise. To my oxygen-starved brain, it seemed as if we would never get there.

As we struggled upward, the slope became steeper with a two-hundred-foot drop straight to crevasses, yawning gashes in the ice, ready to swallow whatever fell from above. Seven Tibetans of the mountain Sherpa tribe, acting as porters for the British, had died here in an avalanche in 1922. The snow is firm, unlikely to give way, I thought to myself. It was wishful thinking, since there is no way to reckon the dangers of Everest.

I lost sight of Woody over the rise of the slope. Now impatient, I unroped and, after twenty minutes of climbing, passed Hans Peter hunched sluglike in his down parka, heaving for breath, too tired to look up. A half hour later I came up behind Norman, who turned slowly to me, peered out from behind his dark goggles, caught a breath as if to speak, shook his head, and turned back to the snow slope. When I reached the front end of the rope, I found Woody was gone. My brain, functioning at the level of a reptile's, failed to understand the danger. For the three weeks since we had left base camp in Nepal, we had worked as a team. Now, in anticipation of arriving at the North Col, two of us had unroped. If one of us slipped, there was no rope anchor, no belay, to brake the plunge to the ice cliffs below.

A star burst of aluminum pitons, the spikes used for anchoring the rope, lay sprawled carelessly on the snow. Woody should have told Norman and Hans Peter that they were off belay. I continued upward.

When I stepped onto the ice saddle of the North Col, Woody greeted me with a bear hug, "We made it, Roger—the North Col!" His smile cracked through old sunburn scabs and layers of glacier cream, and the reflection of the blazing sun slid toward the black shadow of the North Peak in his goggles.

"Goddamn, we can see all the way back to Nepal. The North Col! Twenty-three thousand one hundred and thirty feet on Mount Everest!" He spun around, dancing, arms in the air like Zorba the Greek.

Wordless, I dropped my pack to the snow and flopped down on it, gasping for breath. I squirmed out of the shoulder straps and looked up at the summit of Everest, six thousand feet above us. It seemed close enough to reach out and touch—an easy day's hike in the White Mountains of New Hampshire. By tomorrow we could be at the band of yellow rock where the English mountaineer George Leigh Mallory had disappeared in 1924. On the other hand, no one had ever made it to the top from here, and the slopes above were haunted with the ghosts of dead climbers.

I had never expected to make it to the North Col. I had never shared Woody's dream of being the first to climb the North Face of Everest without artificial oxygen. I was along for the adventure of trekking in the Himalayas and to serve as a porter to set up a few high-altitude camps. But as soon as one camp had been set up, there had been another, and then another. Woody needed my help, not only as a packer, but he had come to rely on me as a route finder, so I'd kept climbing. I had told my father I would stay at base camp, which was twenty miles of glaciers and rock ridges away as the raven flies.

I stood up, stretched, and rubbed my aching shoulders. Below to the north I could see the East Rongbuk Glacier, the route the classic British expeditions and the Chinese in 1960 had used to approach Everest from Tibet. Below to the south, I could see the West

Rongbuk Glacier, the route over which we had donkeyed hundreds of pounds of supplies from base camp in Nepal. Gray clouds reached through the mountain passes like the claws of an immense beast climbing the glaciers from the south. If the monsoon snows broke over the range before we returned to base camp, our route would be covered with wet snow hiding the crevasse mouths. We would have to wallow back to base camp in hip-deep snow with little chance of finding our food caches.

Yes, I felt good, but it was not in my nature to dance and spin like Woody. We were in a perilous place, with no chance of rescue. Neither our families nor the authorities knew we were on Everest. I pulled off my down-filled mitts, fished my camera out of the pack, and took Woody's picture against the summit. I ate some chocolate and tried to regain energy.

Shadows crept down the West Rongbuk Glacier and joined those advancing up the East Rongbuk. The tide of night surged into the lower valleys and drew down the warmth from the sunlit slopes around us. In an instant, the gathering shadows tipped the scales of heat and mass; the wind picked up and funneled with supernatural determination through the icy gap of the North Col. A raven flying frantically sidewise, rose up, and was gone before I realized it was there. Strange, a raven at this altitude. I reviewed the scene in my mind's eye. The raven was still there.

"Rog, are you all right?" Woody waved his ice ax in front of my eyes.

"Woody, we've got another load." I spoke for the first time, choking on the cold air.

"What? You can't be serious. Why?"

"The second tent is down there. The wind is picking up. We can't survive the night without it." I stood up and emptied my pack.

"That's no problem. Let's go get it." Woody smiled broadly and dumped the contents of his pack on the snow. He bounded down the slope and vaulted over his ice ax.

I've skied down slopes steeper than this, I told myself, and bounded off after him. As I passed above Norman, my feet slipped out from under me. I slid several meters before piling into Norman, who grunted, checked his ice ax, and squared his stance against my impact.

Dazed, I pushed myself up, faced into the slope, and kicked

into the footholds with renewed caution. The snow was freezing fast. Chunks of frozen snow clinked beneath my boots, skidded down the slope, picked up speed, and bounced off the ice cliffs below. Through my legs, I saw the purple shadows creeping through hummocks on the glacier two thousand feet below. Although the snow around me was sunlit, the warmth was draining away like a thick mist. A crust of ice armored the slope and turned back the last rays of the sun.

I heard Woody yodeling wildly out of tune and pitch.

Growing anxious, I yelled after him, "Woody, stop! I need a belay!"

There was no reply, only the sound of the wind in the hood of my parka and on the ice. As I waited for an answer, the summit turned pale red and faded like a candle going out. Now that shadows shrouded the cliffs, my mind was left to its own constructions. The slope seemed steeper, more exposed, and more precarious. All my premonitions and dreams of free-falling through the night knotted in my chest. I wasn't afraid of hitting the ice below. It was the downward acceleration, that could tear me apart.

I started inching along on the two points of the crampons, the steel spikes lashed to my boots. I felt tiny in the limitless bowl of darkness, insignificant before the immensity of cold that was draining the last of my energy from me. Without food or oxygen, my metabolism could not fight off the cold. I craved the shelter of the tent, the warmth of a purring cookstove. Falling boulders, pried loose from high ridges around us, screamed like incoming artillery.

"Woody," I yelled into the night.

"What?" he replied, surprisingly close. "What's the matter?"

"The slope is slippery. We need to rope up."

"I'm at the cache. Hurry up!"

"Hold on. I'm coming." I inched down the last twenty meters to the cache and rummaged through the duffel bag. I handed the tent to Woody, stuffed food bags into my pack, heaved it onto my back, and uncoiled a length of gold nylon rope.

"Do you really think we need this?" he grumbled. "We've already been over the slope twice without it."

Unbelievable, I thought. He had taught me to use the rope that made mountain climbing a skill instead of a game of chance or the expression of a death wish, and now *he* didn't want to rope up. He begrudgingly tied one end of the rope around his waist.

I tied the other end around my waist. The wind on the summit had kicked up a plume of spindrift snow that glowed in the twilight like a comet.

As we started up, the rope in front of me draped in loops on the ice; Woody was out of sight over the rise of the night-bound cliff. I could barely see our old footsteps on the ice wall. I moved slowly, cautiously, kicking into each foothold. I stopped often to rest, studied the slope ahead, and watched the planet Venus bob in and out of the watery sky lapping against the mountain ranges. "You have me on belay?" I yelled.

"Yeah, sure; stop worrying," Woody replied gruffly.

We climbed in tandem for a half hour with the rope loose between us. The black mass of Everest slid below the tail of Scorpio, balanced like a supine question mark; Antares blazed above the summit like a bonfire. I thought that by now Woody would be at the belay station. He'd have his ice ax shoved into snow up to the handle, the rope around it. He'd be taking up the slack as I climbed toward him.

Instead, the rope disappeared into the darkness below me.

"Woody?"

No answer.

"Woody," I yelled again. The darkness intensified the depth of the abyss below me. My legs trembled with the fatigue of standing on the steel points.

"What is it?" His voice was carried off by the wind.

"Belay!" I was on the section where I had slid on the way down. It seemed impossible to catch my breath. I was losing energy fast.

"Okay!" His voice seemed louder. "We're almost there."

I saw his dark silhouette hunched against the ice. I stepped out sideways onto an ice rib, dug in my crampon, and started to shift my weight. The crampon gave way, and I careened down the ice slope with terrifying speed.

I grabbed at the loop of rope accelerating past me. I rolled over and over, trying to push my ice ax into the hard ice. I scraped, kicked, and flayed with my crampons. Nothing held. The rope snapped around me and squashed my ribs. For an instant I thought that Woody had caught me.

Then everything gave way. Reflections of stars rushed by like tracer bullets. Shadows swept over me in a rage. A cry began in the base of my spine. Even though there was no sense to it, no help coming, nothing to stop the fall, I yelled and screamed. I would die when I hit the ice below. As soon as I had that thought, my guts and heart pushed upward like the floor of a falling elevator. With excruciating anxiety, like that of a child torn from its mother's womb, my soul ripped free.

Then the strangest thing happened. I shot off into starless space, floated free in zero gravity, and watched my body, as if in slow motion, tumble over the ice cliffs below. I perched on the cusp of time, where, like a water drop between watersheds, I could choose between worlds. I could see in all directions at once, not with the seeing of eyes but the seeing of dreams. I felt no fear and no cold; space seemed to shrink around me, or perhaps I expanded to it. At any rate, I was no longer afraid of the emptiness below me. A great warmth and euphoria overtook me, as if I were immersed in a tropical sea the same temperature and mood as the rest of my being. I thought, *Here you are about to die and you feel wonderful—you are so weird!* As soon as I had that thought, I dropped onto a snowbank, and with furious kicking and falling ice blocks, Woody landed beside me.

"What are you doing here?" I asked.

I later learned that Woody thought that my question showed I had lost my mind, that I had hit my head, suffered a concussion. What I'd really meant to say was, *Why the hell didn't you have me on belay?*

The historians of Mount Everest, who refer to our exploit as the Sayre Expedition, claim that it is impossible to survive a night in the sub-zero temperatures without a tent on the slopes of Everest. Not only did Woody and I survive the hundred-and-eighty-foot fall, but we survived the night huddled together, wrapped in the nylon tent shell. The ledge where we had landed was not wide enough to pitch a tent. The historians believe we were incredibly lucky. They don't know the half of it.

At the time, I was merely amazed and grateful that we had survived. The next day we climbed back up to camp on the North Col and rested. Hans Peter, who was more eager to turn back than I, refused to climb any higher, but helped Woody and Norman establish a high camp on the North Face. The following day I started up to the high camp with a load of supplies. I was trying to cover two days' worth of climbing in one, and it was getting late, I had to cache my load and start back. I had descended only a few feet when I heard a mysterious voice cry out. It might have actually been Woody's shout distorted by the wind, but it seemed like a voice from a dream. I turned and climbed back toward their camp.

The snow outside the tent was stained with blood. I yodeled and stuck my head into the tent. Woody, haggard, the sleeve of his parka torn to shreds, told me that he had fallen again and cut open his arm. It was over, we were done, time to head back. The lack of oxygen and the extreme cold had consumed our energy. We could not gain elevation at a rate fast enough to reach the summit in less than ten days. I had no thoughts of defeat or failure but rather felt relieved to be starting on the return.

Some think we were defeated by Woody's game plan to travel light. "We will climb like turtles, carrying everything on our backs," he had declared. Financing the expedition out of his own pocket for $12,411.59, he could not afford porters or high-altitude oxygen apparatus. More importantly, he had philosophical objections to the nationalistic-combative attitude that the million-dollar expeditions brought to Everest.

Now that the safest routes have been established and the fear of the unknown is reduced with each succeeding expedition, the North Face is routinely climbed without porters or bottled oxygen by solo climbers carrying everything on their own backs. Woody's plan was not so much flawed as ahead of its time.

Many of the conquerors of Everest have become CEOs of sports-clothing corporations, owners of adventure-travel firms, and famous politicians. We approached Everest with no hope for personal gain. Ironically, I gained more than I hoped for. I began a lifelong quest to understand the change of reality that I first experienced on Everest. I delved into the sciences of time and consciousness. I don't think it was an accident that we survived.

Our highest camp was established above twenty-four thousand feet; Woody climbed perhaps another one thousand feet without a pack. After turning back, on the descent to high camp, Woody made a misjudgment typical of oxygen depravation. Deciding to slide down on his butt, he accelerated out of control, tore through a boulder field, and would have shot off the ridge to the East Rongbuk Glacier below had he not grabbed the tent flap as he flew past. That was when I heard the mysterious voice calling me.

I offered to help Woody and Norman down that night, but

they wanted rest, so I descended to the lower camp with an empty pack. The next day Woody and Norman, unable to carry their packs, yelled down to us for help. Hans Peter and I climbed up and packed them down to the North Col. They were in a desperate state, unable to walk, stumbling around, falling asleep in the snow. It was then that I realized we were in a fight for our lives. We would be lucky to make it back to base camp.

The next morning, Woody, barely able to walk, started down the ice cliffs below the North Col. Norman belayed him by tying lengths of rope together. The precise moment that Norman let go of the rope to tie on a new segment, Woody fell. When we heard Norman's muffled gasp, Hans Peter and I looked up in shock. The belay rope, whipping around the ice ax in front of Norman, slithered out of sight over the ice. Woody was gone. After yelling for fifteen minutes with no reply, Hans Peter started down to check on him.

Norman and I waited at the North Col for Hans Peter to return with news of Woody. It seemed unlikely that Woody had survived the fall. Finally at dusk, when Hans Peter hadn't returned, we started down ourselves. I tried to drag Hans Peter's pack behind me, but it almost pulled me off the mountain, so I was forced to leave it behind. Norman and I were soon socked in by night and forced to tie ourselves to an ice cliff. We survived another night without a tent.

The next morning, we heard Hans Peter's muffled shout from below. "Roger, we have to go to Tibet for help."

I looked down and saw Hans Peter, dwarfed, at the base of the cliffs. "Where's Woody?" I yelled back.

"I found him. He's okay. We have to go to Tibet." Hans Peter took off across the snowfield at the top of the East Rongbuk Glacier.

Norman and I spent the day working our way to the base of the cliffs. In the late afternoon we found Woody asleep on the snow. We pitched the tents and Norman helped Woody inside.

When I awoke the next morning, Hans Peter's side of the tent floor was ice cold. After cooking a breakfast of tea and oatmeal, I tried to roll up my air mattress but, without the strength to squeeze out the last of the air, I collapsed on top of it and gasped for breath. It was unusually quiet, and even the flapping of the tents had stopped.

"Woody, wake up." I heard Norman's hoarse voice in the tent next to mine. "You're knocking the stove over."

Eventually I stuffed the air mattress into the pack and stopped again to rest. I looked around at the stove, pot, and cup in the cooking alcove, and rolled them toward me with my foot.

"Sorry, Norman. Can't help it," Woody whispered in a distant voice. "Sleep, jus' let me sleep."

Stopping for breath several times, I tucked my sleeping bag into the top of the pack and fumbled to secure the straps.

"Get your boots on," said Norman.

I had slept with my boots on; the laces had been too frozen to untie the previous night. I fumbled around for my goggles and wool hat.

"Oh, God, we're not going to make it." Woody sighed deeply.

"Shut up, Woody!" Norman's voice trembled with exasperation.

I jostled the pack through the tent flaps and followed it out into the glaring whiteness. I adjusted my goggles and studied the terrain, an amphitheater surrounded on all sides by ridges and peaks. Propping myself up on my ice ax, I paused for a few deep breaths. From below the horizon, the sun was projecting an ultraviolet shadow of Mount Everest onto the ceiling of pink cirrus clouds. I went over the route in my mind: twenty miles to the Nup La, the glaciated pass at the Tibet-Nepal border, then down the Ngo Jumbo icefall to base camp, and then ten miles to the Sherpa village of Khumjung. With the fixed ropes we had already installed, we should reach safety in three days, unless the monsoon ripped through the mountains. Woody's boots emerged from the tent next to me and then they lay still. Could Woody even walk?

"Don't lie there like a bearded fetus. Out! Outside the tent!" Norman scolded.

Woody crawled out backward, dragged his pack after him, and slouched over it in front of the tent. I stood up, using the ice ax like a cane, hobbled toward him, and waved my arm above him. "Hey, Woody!"

He squinted up through glazed eyes framed by deep shadows and scars, lifting his arms as protection from the light. He's in rough shape, I told myself; thank God I don't feel as bad as he looks. In fact, I didn't have feelings. My neural net had frozen up—no thoughts of sorrow or fear. Eat! Sleep! Trudge to safety! Feel the emotions later; pack up the tent now. The barren emotional landscape of my brain was as foreign to me as the ramparts of snow and ice around us.

"Oh, Roger, it's you?" Woody mumbled, surprised. He let his arm drop. "How you doing?"

"Fine, how *you* doing?" I squatted down next to him.

His eyes went dead and cold. His head dropped; he was sinking into thought. Then he looked up slowly. "Roger, I saw it up there." His eyes flared with the light of remembrance. "It's up there." He waved vaguely toward the North Face. Then the spark in his eyes went out. He slouched over and pulled his scarf over his head.

"Woody, what?" I shook him by the shoulder. "What's up there?"

"Yeah." He looked up slowly. "Walls, stone walls, and vineyards; it's incredible. Go up and take a look."

I thought of Frank Smythe's mirages, the strange black floating objects he saw at 27,200 feet on the North Face in 1933. Perhaps Woody had seen something similar. On the other hand, perhaps he was hallucinating. For that matter, perhaps Smythe had been hallucinating.

"I'll go up and take a look, but let's get you back to base camp in Nepal first." I took a deep breath and stood up.

"Nepal?" He squinted into the distance, searching the horizon for memories.

"Yeah, Nepal, home! Let's get you home." I pointed to the range of mountains to the west.

"Sure, Rog, thanks," he whispered, out of breath, and curled up on his scarf and fell asleep.

There was no point to hope or despair—emotions would only waste energy—but Woody had to get better. If he couldn't walk, or if the monsoon snows blocked the way back to Nepal, we would have to seek help in Tibet.

If we ran into serious climbing, Woody was essential. I could find the route, but Woody was the climber, the one who relished the danger and thrill of leading impossible routes. I turned back to my tent and kicked at the piton that anchored the guy line.

"We're almost ready to go!" Norman pushed through the tent flaps, pounding Woody on the back.

I finally got the piton loose, collapsed the tent, folded up the poles, and, sprawling on my hands and knees, rolled up the tent.

Woody looked up. "I can't, Norman." He shook his head slowly. "Leave me. I can't carry a pack. I'll be fine. . . ."

I shouldered my pack and leaned on my ice ax, gasping for breath. I wasn't surprised that Woody wanted to stay behind. I had had the same feelings during the night out on the North Col; somehow, Everest seemed like a good place to die.

"Don't be stupid; you'll never see Edith or the kids again if you stay here." Norman, blue eyes swollen and bloodshot, looked at me for the first time with a frown of worry. He took a deep breath. "You carry his stuff!"

Norman's officious tone brought up the first emotion I'd felt in weeks, *resentment.* I'd rather die than be told what to do. Even though Hans Peter and I had been recruited as packers, we had come to think of ourselves as equal members of the expedition. Besides, Norman was wrong; first we had to set up a camp at a lower elevation; then we could relay back for Woody.

I hefted my pack to dramatize its heaviness. "Can't. Hans Peter's stuff." I turned toward the Chinese camp, leaving Norman to pack up the other tent.

"Goddamn it, don't you leave Woody here!" Norman yelled after me.

I wasn't leaving Woody; but I needed to find Hans Peter. I kicked at the surface of the glacier with my toe. The wind had veneered the blue ice with hard-packed snow. There was no need to put on crampons; I'd be getting off the ice soon anyway. After trudging for thirty minutes, I turned and looked back at the cliffs of ice hanging from the North Col. Diminutive against the mass of Everest, Norman, in his blue windbreaker, was shuffling forward, head bowed under his pack. Behind him Woody, in his torn orange windbreaker, was dragging his pack on the ice by a rope. When it got stuck in a drift, he tugged at it as if it were a recalcitrant burro, then, shoulders heaving, flopped on the snow.

I continued down the glacier, staring at the snow surface as I walked. Patches of blue ice showed through a miniature landscape of windblown dunes, streamers, and drifts. I felt as if I were looking down onto a snowy landscape of mountain ranges, valleys, watersheds, and river tributaries. With giant steps I skimmed over a microcosm of continents, mountain ranges, and Arctic wastes.

A ghostly yellow shape in the ice startled me. I felt a strange compulsion to avoid stepping on it, as if it were a dead animal on a country road. Throwing my arms up for balance, I stumbled, hopped, skirted the object, stopped a few steps away, and stared at the shape embedded in the pale ice. It must have been there for years—Maurice Wilson's tent, I thought to myself. It had been frozen in the ice since the day he died, May 31, 1934. Eric Shipton, the leader of the British expedition the following year, had buried the tent in a crevasse, along with Wilson's body, but the body and tent "reincarnated" during the 1960 Chinese expedition. The Chinese had thrown Wilson back into a crevasse, but here he was, at the surface again, two years later.

I took a closer look at the tent and saw a bone sticking out from one corner. Gross, I thought. It was the first corpse I had ever seen. I had avoided death all my life, but on the slopes of Everest it surrounded me: Mallory, his companion Sandy Irvine, the seven Sherpas, Maurice Wilson, and soon, possibly, Woody. The "Mother Goddess," Mount Everest, was leading me into a lifelong study of death, a study that forced me to change my idea of what it meant to be alive.

I felt respect and empathy for Maurice Wilson. If our expedition is mentioned at all in the histories of Mount Everest, it is recounted in the same chapter as Maurice Wilson's adventure, under a title such as "Outsiders, Fanatics, and Outlaws." Often, we are left out and Wilson is given his own chapter, with a title like "The Mad Yorkshire Man." It seemed ironic to me that Maurice Wilson was considered less sane than other Everest climbers. Quite simply, in comparison to the normal range of sanity in civilized society, Everest climbers are off scale, at the far end of the spectrum. However, Maurice Wilson, searching for a rarefied, spiritual state of being, has been labeled "mad," whereas those climbers who gained fame and fortune by reaching the summit, by triumphing over, conquering, the mountain, are justified in the eyes of society. After all, fame and fortune are sane objectives.

I paused for a minute of silence, then turned and studied the route ahead. The ice field funneled into the upper reaches of the East Rongbuk Glacier. Like a snowplow, the glacier had piled up moraines, long ridges of dirt and boulders, on either side. I headed for the moraine on the south side, and when I reached it, the fresh soil seemed joyfully warm and, in comparison to the ice we had been climbing on, full of life, even though nothing but bacteria and lichens grew there.

Indeed, several inches below the surface there was a core of ice. At every step the slate fragments slid under my feet. I stumbled, recovered my balance, and stumbled again, sprawling facedown on the moraine. It felt great to be lying on dirt instead of ice. The radiant heat penetrated my windbreaker and soothed my nerves. Squirming out of my pack, I rolled over on my back and watched Everest ride the wake of clouds streaming toward Tibet. We had survived the North Face. In comparison, the route back to Nepal would be easy. I pushed myself up on my ice ax, and adjusted my pack on my shoulders. In the distance, I could make out the low stone walls of the Chinese campsite, tracing over the moraine like the ruins of a fortress.

I stumbled down the moraine late in the afternoon, and as I entered the Chinese camp, I felt—without actually seeing him—Hans Peter stir in one of the stone-walled enclosures. "Hans, there you are! I thought I'd be chasing you all the way to Lhasa."

He jumped up startled. "Oh! Roger," he replied in his Swiss-German accent. "No, no, I waited for you."

I dropped my pack and pulled out the tent. The aluminum tent poles rattled as I unrolled it on a flat patch of sand. "I brought your sleeping bag and mattress but had to leave your pack. It almost pulled me off." We were covered by shadows, but the snow-covered peaks around us still blazed in sunlight.

"Thanks. I nearly froze last night." He stood up and rubbed the cold out of his limbs. He removed his mitts, blew on his hands, and picked up the rear tent poles. Working the front of the tent, I studied him as he fit the aluminum poles together. Although his face was haggard, his blue eyes still glowed with life. It was good to be pitching the tent on earth instead of ice. He looked at me and smiled weakly.

"Hans"—I paused, searching for the right words—"I don't think we should go to Tibet."

His smile dropped, his jawline tightened, and he looked down at the tent. "Why? We don't owe these guys anything."

"Because—" I started to explain.

"Woody tried to pull me off last time he fell," he interrupted.

"Really?" I was surprised to hear this.

"Yes," he replied sternly. "And I was trying to help Woody."

"What happened?" I fed the front poles through their sleeves.

"Well, Woody slid about a hundred feet and ended up in a crevasse." Hans Peter shrugged.

"A crevasse, Christ! So you helped him out?" I looked at Hans Peter in disbelief.

"By the time I got down, he had climbed out of the crevasse by himself and traversed out onto the steep ice slope. Incredible! Tough guy." Hans Peter shook his head as he fed the rear poles through their sleeves.

"And then?"

"He kept yelling, 'Hans, help me.' "

"Did you?"

"I climbed down until I found good footing. By that time, Woody was above me but couldn't hold on, and he said, 'Okay, what's one more fall,' and let go. As he slid by me, he reached out and yelled, 'You're coming with me' "

"Perhaps he was pissed at you. Perhaps he thought you were the one who dropped the belay or that you weren't helping him."

"*Ja,* perhaps so, but he's also crazy. Anyway he slid another four hundred feet. I climbed down to him. We spent the night in a snowbank huddled around a stove. *Scheisse!*" Hans Peter fell quiet, then drove a piton into the sand, hooked the rear poles together, fastened the guy line to the piton, and pulled up the rear of the tent.

"Woody paid your way," I said after rigging up the front of the tent. "You owe him."

"Well, then. We will get help in Tibet." Hans Peter, loved mountains and the people who lived in them. Ever since reading Heinrich Harrier's *Seven Years in Tibet,* he had fantasized about trekking through Tibet to Lhasa.

"That might work for you, you're Swiss, but Woody's afraid we'll be arrested, that they'll lock him up forever, even execute him." No Americans had been in Tibet since the Chinese occupation of 1950.

Hans Peter's ears flushed red with anger, and he stomped the guy-line piton and shouted, "I want some hot tea, all right!"

He got the stove going in a corner of a stone wall, melted down some snow, and threw in the tea bags. We leaned over the steaming cups like Mongolian herdsmen. It hadn't occurred to me that Woody had been crazy all along.

"How is Woody?" Hans Peter asked at last. "When I left him, he couldn't walk."

"He keeps falling asleep and hallucinating. Did he hit his head?" I asked.

"I don't think so, but look at how many times he's fallen. What is it? Four times . . . thousands of feet? At this altitude, without oxygen. Must be a record, might be brain damage from lack of oxygen. Not that he didn't have brain damage before the climb." He winced and took a sip of tea.

"Well, we're lucky." I put down my cup and looked directly at Hans Peter. "We're off the mountain; dozens of climbers have died here."

"*Ja,* Woody's lucky, all right. But he's also crazy." Hans Peter stared directly at me, unflinching.

I considered it our duty to wait for Woody. When, two years earlier at the age of nineteen, I was in Antarctica with Professor Robert L. Nichols, he posed the following question: "You and another climber are crossing a glacier roped together. The other guy falls into a crevasse and is knocked unconscious. What do you do—cut the rope and go for help, or stay with the injured man and face death?" The only acceptable answer to Professor Nichols and, I dare say, every explorer who set forth in the first half of the twentieth century was, "You stay with the downed man, no matter what." Not that anybody ever did that, but it was the attitude, the heroic sense of duty that mattered.

The world seems more cynical now; it's every man for himself. Climbers and paying clients wander all over Everest without concern for each other. I once idolized the heroes and pioneers of the American frontier and the explorers of the polar regions. Exploration was an end in itself. Norman resented my idealism, my gung-ho attitude, and claimed I was a danger to the expedition. Would I cut the rope, or stay and die with an injured partner? I'm not sure anymore; depends on the partner. But on Everest, I was sure I would stay with Woody whether Hans Peter did or not.

"Woody's old, but he's a tough old guy." I drained my tea cup. "Look, Hans, I agree with you. He's not the hero I thought he was. But if we get him to a lower elevation, he might get better. We've got to stick together."

"All right! All right!" yelled Hans Peter, throwing his arms in the air. "God looks after fools but leaves it up to us to pack them out."

I breathed a sigh of relief when he threw down his cup and stormed up the moraine, leaving me by myself.

I looked around at the ramparts of ice and rock. As the shadows at the base of the mountains thickened, a chill of loneliness went through me from the inside outward. Another cold night in the Himalayas. Night did not fall in these mountains; instead the shadows climbed upward, sliding across barren cliffs like supernatural beings. Approaching the summits, they leapt from ridge to ridge with increasing speed and agility, then devoured the last morsels of light from tips of the highest peaks, and closed the vault of endless darkness.

In the blink of an eye, the vastness of space drew the energy and hope out of me. Even the emerging stars gave no comfort. At that altitude they did not twinkle, but stared, cold, steady, and distant. I was alone in the bottom of an icy ocean of space too bleak even for ghosts. Tibetan Buddhists advise the dying to let go of earthly possessions, to focus on the white light of the afterlife. In the Himalayas, the purgatory of cold and loneliness burns away the attachments that bind ghosts to earth. Like the shadows at twilight, the souls of the dead shoot straight from the snow-clad mountaintops through the vault of the heavens without thinking twice.

I crawled into the tent, curled into my sleeping bag, struck a match, and lit the stove. The familiar smells of sulfur and butane, the sound of the blue flame purring like a cat, gave me comfort. But I couldn't shake the image in my mind that only the thin membrane of the nylon tent sheltered me from the mountains of shadows that stretched all the way back to the big bang. I dozed off but woke up when I heard voices outside.

"Oh, God, Oh, God, I never thought I'd make it." Woody's voice trembled with fatigue.

Hans Peter unzipped the door of the tent, shined his flashlight through the flaps, and stuck his head in. "I've got Woody—he just barely made it—and look what else I've got."

Two rusty number-ten cans sloshed across the floor of the tent. Liquid water! I couldn't believe my ears. As soon as Hans Peter punctured the first can with the opener, the pungent smell of dill exploded into the tent. Who would have believed it, whole cucumber pickles! The second can contained chicken in sloshing broth. God, the Chinese must have attacked the mountain with an army of coolies. Imagine carrying water up a mountain! No meat bars and dehydrated vegetables for them. We gnawed the bones clean. At last I felt warm and sociable and looked forward to a good night's sleep. Yes, we had truly escaped the sliding gales of the upper atmosphere. Good-bye, stratosphere; hello, troposphere, we're back.

The next morning I woke to the sound of Woody thrashing, and shouting, "Norman! Norman! The Chinese. I see their lights."

"Wake up, Woody!" yelled Norman. "You're having a nightmare."

Even though he had just been dreaming, Woody's paranoia was justified. With no physical obstacles between us and Tibet, a Chinese border patrol could easily make its way to our camp. Woody's fear was that the Chinese would incarcerate him, the grandson of American president Woodrow Wilson, as a spy.

As we packed up, Woody lay in a fetal position on a pavement of slate slabs.

"Front and center over here." Norman had taken command of the expedition, and even though he was truly concerned for his longtime friend, he also believed Hans Peter and I were subordinates, porters brought to help him and Woody to the summit of Mount Everest. Woody, as leader, had refused to let Norman treat us as coolies. Hans Peter and I, ignoring him, continued to strike our tent.

" 'You're in the Army now!' " I sang softly to Hans Peter.

"You boys were paid to come here as logistical support," Norman lectured. "It is now time to perform the function for which you were paid. Woody needs you to carry his pack."

Hans Peter kicked the piton out so hard the guy line wrapped several times around his foot. Swearing in German, he struggled to unwind himself.

I shrugged my shoulders and pleaded my case with palms turned to heaven. "I'm loaded with food in case we miss the East Rongbuk cache."

"Dump it!" commanded Norman. "Woody's stuff is more important."

I unloaded canned Chinese crabmeat and whole chicken and substituted Woody's sixteen-millimeter movie camera, exposed reels of film, and a stove with butane cartridges. Hans Peter strapped a tent and our last climbing rope to the outside of Woody's pack and shouldered it.

I was distracted by two large ravens, known in Tibet as goraks, flying low, fighting the stiff breeze like seagulls landing on a pier in an offshore wind. One raven stuck his head inside an empty can of condensed milk, flipped it over his head, and then looked at me as if to say, "Good trick, eh?"

The other raven, hopping on one leg with the opposite wing extended for balance, focused on a lump of oatmeal on Woody's beard. The bird cocked its head, flapped its wings, and dove for the oatmeal. Woody woke up with a start. "Norman! Norman! What's going on?" The raven landed adroitly, bolted down the oatmeal all in one motion, and looked at me to see if I had noticed.

"We're going. Take this." Norman handed Woody a Chinese bamboo pole and helped him up.

I shouldered my pack and grabbed my ice ax. Woody walked stiffly a few steps, then stopped and rested.

"You guys go ahead; find the cache," said Norman.

Hans Peter and I started down the moraines. On the trek in, we had cached a duffel bag of food near the base of the East Rongbuk Glacier. We had also left a bag of sugar and tea at the main Rongbuk crossing, and a large cache of food and climbing equipment at the Nup La. On dangerous or difficult sections of the route, we had left fixed ropes secured to pitons and ice screws, especially on the Ngo Jumbo icefall above base camp. The worst was behind us, I thought; we would be in base camp in a few days. I turned north, following the edge of the glacier.

As I walked deeper into the glacially carved canyons, I enjoyed watching the shapes of the cliffs, boulders, crags, and ridges move around me and unfold into new patterns. The distant landscape, two-dimensional like a matte painting used in the movies, opened up; ridges swung by like giant doors leading from two-dimensional reality into the world of three dimensions. The act of walking created the third dimension; sometimes I imagined that was my work on Earth. Walking in the mountains enlightened my psyche and clarified my mind. I enjoyed picking my own route, making my own choices, watching the mountain faces pass by. I cherished the freedom of trekking in the wild places of Earth.

I was not alone in my enjoyment of walking in the mountains. Many ancient peoples believed in the power of ritual walking. With stone markers, they laid out courses to be walked only with the greatest intention and ceremonial respect. In Tibet, people believe walking long distances to holy places will purify their bad deeds. The more difficult the journey, the deeper the purification.

Walking west over the moraines, I felt my strength return. Hans Peter and I were soon far ahead of Norman and Woody. I turned to watch an avalanche pour down the North Face. The crashing sounds were strangely out of sync with the tumbling blocks, like a badly dubbed kung-fu movie. Over those great distances, the light arrived before the sound. I watched in fascination as the last blocks ground themselves to bits and drained through the black cracks in the cliffs above the glacier.

No living thing can survive for long in the loins of the Himalayas. The earth is moving too fast for plant or animal to gain a toehold. The awesome walls of rock and ice are covered with dust and debris, like cliffs along a newly blasted freeway. Rocks, pried loose from the crags by weathering and erosion, bounce off the ramparts with shrapnel-like explosions and collect in mile-high scree aprons at the base of the cliffs. Like conveyor belts, the glaciers pick up the fragments and carry them to the lowlands. A skittering slab of shale can set off a chain reaction of debris and dislodge house-size blocks of the mountainside.

Even though it's continuously being torn down, Mount Everest grows in elevation. The Mother Goddess of the World thrusts new crust from inside the earth toward the rarefied heights. I felt like the slow-motion version of a Marvel Comics action hero riding the mountain up as it crumbled under my feet, ice ax thrust high, lightning blazing from the iron tip.

Harm done to the beings roaming the base of the cliffs was unintentional but a serious possibility since, preoccupied with giving birth to continents, the Mother Goddess hardly noticed our presence. No longer in danger of falling off the mountain, now we had to worry about the mountain falling on us.

Hans Peter and I reached the East Rongbuk corner by the time the shadows had slanted up the valley walls. Any loose rock would be frozen in place until the sun hit the slopes on tomorrow's morning. Our immediate concern was to find the food cache we had left in a duffel bag under a boulder. The boulders all looked the same, and the fresh ones, just fallen from the cliffs above, had totally changed the scene. After an hour of searching we still hadn't found the cache. Then a movement caught the corner of my eye and I turned to see a white plastic bag roll along the moraine like tumbleweed fleeing the wind.

"Oh! Oh! Did you see that?" asked Hans Peter.

"Yeah, right. Oh! Oh!" I joined in. We traced the path of the bag back to a patch of ground strewn with white plastic bags puffing and deflating in the breeze, each with a hole ripped in it, the contents gone.

We had sealed the food in the white plastic bags inside a duffel bag and covered the whole thing with stones. The stones were now thrown aside; the duffel bag was shot through with holes. The food was gone, not one almond, raisin, or bit of chocolate.

"Ravens," said Hans Peter.

"Yeah, but look at this." I pointed to a spare camera ten feet from the cache. "Can a raven carry something that heavy?"

If they did carry the camera, they were pretty amazing ravens. And why would they even bother with a camera? Dejected by the destruction of the food cache, we sadly looked for a spot to pitch the tents. The only flat place was under a steep slope of loose scree and boulders that climbed straight to the black cliffs above.

I saw a shape moving on the other side of the East Rongbuk Glacier. "Maybe a yeti," I said, pointing out the moving shape to Hans Peter. We had seen footprints on the way to base camp.

"*Ja,* or maybe a mountain goat or a bear. Who knows?" Hans Peter scanned the moraines for another view of the creature, but it had disappeared behind a boulder.

In my mind's eye the creature walked upright and had a somewhat pointed head. "No, it seems more like a yeti."

When we crawled into the tent, snowflakes pelted the orange nylon, leaving wet spots that grew on the weave like alien lifeforms. Our tents, designed for snow and ice, weren't waterproof.

"Monsoon," I said, feeling the cold humidity through my parka.

"*Ja,* it's wet; it came from India." Hans Peter pulled the stove out of his pack.

"We'll decide what to do, which way to go, at the Rongbuk crossing, if we ever make it." I was beginning to feel that as soon as things were going smoothly, they changed.

We set up the stove, boiled down snow, made our stew of meat bars and dried mashed potatoes, and crawled into our sleeping bags.

Four hours later, well after dark, we heard Norman's voice and saw the glow of his flashlight on the tent fabric. "Roger! Hans! You guys here?"

"Over here!" we shouted.

"What about the cache?" Norman's voice was tired.

"Empty; ravens, goats, yeti, or something," I said. "Hans and I will relay back to the Chinese camp for more food."

There was a moment of silence with only the chorus of faint hisses from the rain of snowflakes.

"No, we'll all go ahead. We have another cache at the main Rongbuk," Norman replied.

"Just tea and sugar," I answered.

"Tea and sugar, that's all we need," replied Norman with renewed authority.

The next morning Hans Peter and I cooked up the last of our oatmeal and hot chocolate and started to break camp. It had stopped snowing, and a light dusting of powder covered the scree slopes, now illuminated in bright sunlight. Woody was sitting on the moraine not far from Norman as he broke down their tent. I heard the faint tinkle of shale and watched a small piece of scree scurry down the slopes. We felt a deep sonic rumbling through the soles of our boots. Then a flurry of free-flying boulders whirred down like incoming artillery.

"Rock slide!" I yelled, rolling into a ball and covering my head with my pack. Hans Peter didn't need to be told; he was already under his pack. Norman groped around for his.

Woody sat up, looked around, and said, "What's going on?" Three boulders whizzed by his face like cannonballs.

Since it was impossible to know which way the boulders would bounce, it was impossible to know which way to run. It was best to hunker down in one spot. Puffs of dust appeared across the top of the slope, and the whole mass gave way. Boulders running loose in front pelted the ground around us. Then, with a consummate roar, the whole mass moved down. The sky darkened. I closed my eyes and waited, choking on the dust.

The commotion died out, a few pebbles bounced off my pack, and the air cleared up. I peered out from under my pack. Woody was sitting in the same position, like a Buddhist monk, in his orange windbreaker.

He squinted at the slopes around him like half-blind Mr. Magoo. "Wow, did you see that?"

The landslide had parted in the middle, diverted to the sides by a small ridge above camp. As Hans Peter had said, Woody was incredibly lucky, a real Magoo Buddha. How many disasters could one person survive on one expedition? The only thing we could do, and it was no great comfort, was shoulder our packs and continue the trek to the main Rongbuk Glacier.

The next day, we arrived at the termination of the main Rongbuk Glacier, where the downward flow of ice was balanced by melting from the sun. Streams poured out from under a layer of boulders and gravel. Weirdly shaped seracs, towering sculptures of ice, had melted out of the network of crevasses behind the terminal moraine.

In this field of sun-sculpted seracs at the junction of the East and main Rongbuk Glaciers, I searched for our cache. This one will be gone too, I thought to myself. My role as route finder, a chore I enjoyed, was to decipher the shapes of the landscape from different angles, to study the way the land spread apart and opened, to interpret the continuous evolution of forms around me. I found the bag of sugar and tea and, not far away, the route with the fixed rope through the seracs. Thank God, I didn't have to cut a new route.

The seracs, surrounding us like the monoliths at Stonehenge, focused the sunlight on us. We stripped off our windbreakers, shed our down parkas, and stretched out on a patch of toast-warm moraine. Hans Peter soon had a pan of water simmering on the stove, and the four of us huddled over our tea and mulled our fate. We had descended from twenty-four thousand feet to about seventeen thousand feet in six days. It was time to decide. Should we go up the West Rongbuk Glacier to Nepal or go down the main Rongbuk Glacier to Tibet? Each of us stared into our cups as if the answer lay inside. We exchanged glances, then looked back into the tea.

Every afternoon the monsoon clouds advanced into Tibet, and every night they retreated to the plains of Nepal. One of these nights they would not retreat. In warning, a low monsoon cloud passed over us, the pale sun bobbed in and out like a memory, and a sharp chill went through me. We hurried into our down parkas, pulled over our windbreakers, and donned our wool hats. As I shouldered my pack, I looked at Norman and Hans Peter, waiting for one of them to make a decision. They looked blankly back at me.

Woody shuffled over to the pile of equipment, picked up a tent and a climbing rope, and tied them around his neck. He managed a weak smile to me. "Lead on, route tracker, to Nepal." He pointed up the West Rongbuk Glacier with his bamboo pole.

Woody's acknowledgment made me feel great. *Route Tracker,* that's me. Super powers include locating lost caches, avoiding hidden crevasses, recognizing the same shapes from different angles, and avoiding objects about to crash down from above. Yes, I am *Route Tracker,* able to track down lost glaciers, finder of the way through the mists of time, he who turns two dimensions to three. The *Magoo Buddha* is my leader and mentor.

Thank God, he's getting better, I thought to myself. Perhaps he'll be able to lead the icefall. But he had difficulty navigating the loose scree of the West Rongbuk Glacier's moraines. We lost precious days.

Dangerously behind schedule, we made it to the West Rongbuk serac crossing, where we had left fixed lines. I dropped my pack. Woody sat down on a rock to catch his breath. Norman and Hans Peter leaned back against boulders and propped up their rucksacks, wrists crossed at their waists, heads drooped.

I guess it's up to me, I thought. I climbed an ice hummock, searched for the fixed lines, and studied the route. Fog shrouded the tops of the seracs; the ice walls plunged straight into meltwater ponded in the depressions. Since the water was too deep to wade through, we'd have to climb over the seracs.

"How's it look?" asked Hans Peter

"No sign of the lines," I replied. "The seracs are flooded. We've got to climb over them."

Norman dropped his pack with a gesture of disgust. Hans Peter, shaking his head, untied the pitons and crampons from his pack.

As I sat on the moraine lacing my crampons to my boots, I looked at Woody.

"Woody, crampons." I pointed to his boots.

"What? What's happening?" He sat up and looked around.

"You get started; I'll get his crampons on," said Norman.

Me? Without Woody? I thought. Do a lead without Woody? I had done a few minor leads with Woody belaying me, but he was the leader.

"Look at Woody; do you think he's going to lead?" said Norman sarcastically.

"No, I guess not."

"I'll belay you," said Hans Peter.

"Are you crazy? You don't need a belay. There's no more than a fifteen-foot drop," growled Norman.

I had already fallen hundreds of feet on the North Col, so what's another fifteen? I guess that was the attitude I was supposed to have. Who knew for sure if it was only fifteen feet? It could be fifteen or one hundred and fifty. There were no reference points against which to scale the distance. Besides, the fall on the North Col had made me all the more cautious. I shuddered at the memory. I really couldn't cope with another experience like that.

"If I do this route, I'm doing it on belay," I replied.

As soon as I took a few steps, fear overtook me; my adrenal glands went into aftershock and bailed out. It wasn't the prospect of death, it was that of the pain that would precede it. It was the humiliation of being out of control, the unmanly screaming. Furthermore, a fifteen-foot fall into cold water could be more than unpleasant. If I so much as sprained my ankle, it could be the end. We were so close to the Nup La at the Nepal border, why risk an injury now? Why take chances when we didn't have to?

As I cut steps, the chips from my ice ax slid down the steep face. I rested, then stepped onto the new tread, drove in an ice piton, and clipped the rope through a carabiner that would stop me from falling to the bottom of the serac. That is, if someone was manning the belay. The rope, once fixed in place, would also act as a handrail.

After two hours I had advanced about fifty feet. I was exhausted; my breath came hard. My legs started to bounce up and down uncontrollably. The seracs come forward and retreated with my heartbeat. The mist cringed with pending twilight.

"Christ, Roger," yelled Norman, "we could drive a herd of yaks across that. What's the matter?"

"I don't know. Why don't you do it, Norm?"

"That's okay. Carry on." Norman shrugged his shoulders.

Come on, get a grip, I told myself. Climbing is a mental thing. As the saying goes, "Don't look down." Christ, I had already looked down, and I couldn't shake the image in my brain of sliding to the bottom, smashing my ankle, and becoming entombed in the glacier like Maurice Wilson. *Okay, trust the rope.* It would catch me before I smashed my ankle. But then, I couldn't remember the last time anyone was caught on belay on this expedition. What good was the rope if no one was going to hold on to the other end? *Forget the rope; focus on the steps,* yes, *footholds with firm treads, three inches deep.* With sharp crampons, this was not much more dangerous than walking up a ladder. Yes, a *ladder,* that was good. I was walking on a kind of sidewise ladder. It didn't matter how far the drop below was; it didn't matter if anybody held on to the rope. Why should I fall at all? My legs stopped bouncing up and down. I took several deep breaths, stepped across, clipped a carabiner into a piton, fed the rope through, and peered ahead into the swirling mist.

Between the ice cliffs I could see a fleeting shadow of black moraine on the far side. Inspired by the sight, I cut the last dozen steps with new energy, and jumped down onto the gravel. I drove a piton into the last ice cliff and tied off the rope.

"Off belay," I yelled.

I faintly heard Hans Peter's yell swallowed in the mist. "Ooookaaay!"

The others started out one after the other. I sat down on the moraine, head between my legs, gasped for breath, and unlaced my crampons. I heard Hans Peter kicking into the steps. Soon his dark shape loomed in the mist between the seracs. He faced into the cliff, meticulously taking one step at a time and slid the carabiner on his waist harness along the fixed rope. Twenty feet behind him came Woody, carrying my pack, teetering and stopping to catch his breath at every step.

Hans Peter reached the moraine, unclipped, and climbed down. "No going to Tibet now."

"I guess not." I stood up with my crampons in my hand.

Norman was last, and by the time he appeared Woody had reached the moraine, staggered forward, and fallen backward onto the pack. He breathed heavily for a while, then looked slowly at me. "Good lead," he said, and squirmed out of the shoulder straps.

The additional approval from the Magoo Buddha cheered me a bit. It did not occur to me that he might be buttering me up so that I would lead the Ngo Jumbo icefall to base camp. We were simply too brain-dead to have such complicated thoughts.

Norman arrived. "All right, that's over with. Time to pull the rope through."

I had doubled it through the pitons so we could pull on one end and the other end would loop back to us. I tugged on the rope, but it didn't budge. I stood up and tugged on it harder. No dice. We all pulled on it.

"A knot's stuck in a carabiner," I said.

Hans Peter put his hands on his hips and looked to the heavens. "What more could happen?"

"Great! Our last rope," said Norman with bitterness. "How are we going to get through the crevasses?"

"Well, who's going back to untangle the rope?" asked Woody, slumping to the ground.

"My crampons are off. Besides, I'm too fucking tired to climb back and untangle it." I hung my head and stared at the crampons in my hand

"We can't leave it," said Hans Peter. "What if one of us falls into a crevasse? What'll we do, tie our clothes together to make a rope?"

"Nobody's going back. We'll take our chances with the crevasses," said Woody. "We have another rope at the Nup La cache."

"If we get in trouble we'll come back," said Hans Peter.

"Sure, we'll come back," said Norman

"Sure, why not?" I said.

It was late afternoon when we shouldered our packs and headed up the West Rongbuk Glacier like four Old Testament prophets hardened from a diet of wild honey and raw locusts, hair matted, beards ragged, faces emaciated, and, with one exception, carrying Chinese bamboo poles in place of ice axes. We had lost three ice axes, two climbing ropes, one backpack, about forty pounds of food, and, among us, that again in body weight.

Turning back for a last glimpse of Everest, I watched it fade into the pall of the afternoon monsoon. I'd never had illusions of reaching the summit. Like Hans Peter, I believed reaching the summit was irrelevant, perhaps even irreverent. Summiting, especially in the context of conquering, trivializes the mystery. The Mother Goddess grew silent but was not conquered. The clouds closed on Mount Everest. That was the last time I saw the snows of Everest, except in recurrent memories and dreams over a lifetime of trying to understand the experience.

Our trials were hardly over. We were in a whiteout— snow below, snow above. Not able to see more than twenty feet ahead, I stumbled over hummocks, dropped off small precipices, and struggled to keep my comrades in sight.

Woody led, packless, scanning the white expanse for signs:

breaks, cracks, and furrows in the snow. It was like crossing a minefield without a metal detector. He would probe with his bamboo pole, then step cautiously forward. The rest of us, carrying the heavy packs, followed in his footsteps, hoping that the snow bridges would hold the extra weight. I wished I could trust to fate like Woody, but the thought of breaking through to a crevasse made me anxious.

The vague memory of food at the Nup La cache kept us plodding on, hour after hour. We had to make it to that cache. But soon the storm turned into a full-fledged blizzard, and Woody stopped, his beard and eyebrows caked with ice that made him look old and emaciated. None of us looked too good. Norman threw down his pack with a dreary look on his face. We hadn't eaten for two days, and I could not imagine an expedition in a more precarious position. Hans Peter and I plodded through the routine drudgery of setting up camp, went to bed without food, and fell into a sleep without dreams, emptied of reasons for waking up. It was a matter of pure dumb luck, or a miracle, that no one had fallen into a crevasse that day.

When I did awake, I stared at the faint orange glow on top of the dark drifts that were pushing the sides of the tent in. I heard Woody unzip the door of his tent.

"Oh, wow! Beautiful," he muttered.

I pulled on my boots and crawled out. Norman was brushing the snow off his tent. The mountain walls of the Nup La pass disappeared upward into a dense gray cloud layer brushed pink by the rising sun. Golden whirlwinds of ice crystals danced across the white expanse. I brushed the snow off the tent. Hans Peter, dark hollows under his eyes, crawled out, rose into the light of dawn, and scanned for the cache flags.

"Nothing," he said. "It could be under ten feet of snow."

"What if we can't find it?" I asked.

"Perhaps the ravens." Woody shook his head.

"We'll never get down the icefall," Norman said, scanning the horizon with his hand over his eyes.

"We'll go back to the Rongbuk crossing," sighed Hans Peter. The thought of retreating was appalling, but there was no way down the Ngo Jumbo icefall without a rope.

Two whirlwinds crisscrossed the snowfield like children playing tag, traversed in front of us, collided, and burst into a cloud of shimmering ice crystals that floated across the sun like dust floating through a sunbeam and then settled around a small black sliver in the snow.

"There's something," I said. "Perhaps the lip of a crevasse."

"*Ja,* it's something." Hans Peter took off across the snow.

"Hans, the crevasses," I yelled after him.

CHAPTER TWO

DUMB LUCK

Because it is there. The famous British mountaineer George Leigh Mallory, on a lecture tour of America in 1922, gave that response to a reporter who asked why men try to climb Mount Everest. The phrase has taken on the meaning of doing something for its own sake. Was this a tired brush-off of an unanswerable question? Woody and I believed not. To us, Mallory's answer had a deeper meaning connected with the mystery of life, with an eternal feeling that comes in high and holy places. To us "it" was a transcendental state of completeness. My reading of Emerson and Thoreau had convinced me that a heightened transcendental state was part of human nature, even though I had not experienced it directly. Could the "it" Mallory spoke of be the same as the heightened state of the transcendentalists? What was that heightened state? Was it real or imaginary?

Through the ages, legends and songs have advised us to go find "it" on the mountain, and that was our goal—to find IT on the mountain. But we barely spoke of our goal for fear of being labeled mad, like Maurice Wilson.

We approached Mount Everest with a sense of adventure, a need for exploration, a great hope for discovery, and as a test to discover how we would react in extreme conditions. And, for me, Everest was just the beginning.

Woody opened the first door for me. We were drawn to the

mystery of Mount Everest by reports of strange visions, such as the floating black objects reported by Frank Smythe while climbing alone on the North Face in 1933. We were fascinated by those climbers who had tried the impossible.

Foremost among Everest mysteries was that of the disappearance of Mallory and his climbing companion, Sandy Irvine. On the morning of June 6, 1924, George Leigh Mallory, age thirty-nine, and Sandy Irvine, age twenty-two, left the North Col for camps higher on the North Face. Two days later, carrying bottled oxygen, they set off for the summit. Noel Odell, a geologist climbing toward high camp, looked up through a sudden clearing in the atmosphere and saw the silhouettes of two figures climbing the rock step on the summit ridge. Odell concluded that he had seen Mallory and Irvine. They were never seen again. What happened to them? Did they reach the summit?

For seventy-five years the disappearance of Mallory and Irvine remained a mystery. According to one account published shortly after their death, the spirit of Irvine contacted a psychic friend of Noel Odell's and gave the following account: While descending from the summit late in the day, Mallory slipped to his death in the darkness. Irvine continued on alone but soon collapsed from fatigue. As he huddled in the bitter cold, the image of Mallory floated before him. "Come on, old chap," Mallory said. "It's time for us to be getting along."

Although I had heard of many psychic visions and predictions, they had always turned out to be wrong. They were too general to be valid or actually made after the fact. I didn't pay attention to them.

On May 1, 1999, four members of the Mallory and Irvine Research Expedition found Mallory's body at nearly twenty-seven thousand feet, right leg broken just above the boot, right elbow either broken or dislocated, rib cage shattered, and forehead smashed in. His sun goggles were in his pocket, indicating he had fallen after dark. The rope around his waist, frayed where it had hung up and snapped on a rock, proved he and Irvine were roped together at the time of the fall. The body of Irvine, found by a Chinese climber in 1975, had been considerably off the line of Mallory's fall, showing Irvine had continued on after Mallory fell.

In light of the new facts, the account of Odell's psychic friend seems remarkably accurate. Mallory fell after dark. He was roped to Irvine. Irvine did continue the descent after Mallory fell. The psychic's account of Mallory's death was published long before his body was found. Could other psychics' observations also be true? My scientific training taught me that extrasensory perception was impossible. How could something that would be discovered only in the future be known in the present?

There are similarities between Mallory's fall and the one Woody and I experienced. Both falls occurred with climbers roped together in the dark on the North Face of Mount Everest. At the age of twenty-one, I was a year younger than Sandy Irvine; at age forty-two, Woody was slightly older than Mallory. Mallory and Irvine were experienced and had the advantage of bottled oxygen and porter support, but they were at the limit of fatigue at an extreme altitude when they fell.

Before our attempt, we thought the North Face of Mount Everest was yet unclimbed. We later found out that three Chinese climbers claimed to have reached the summit in 1960 with bottled oxygen and an army of porters. Counting the Chinese, nine climbers, all using bottled oxygen, had reached the summit of Everest prior to our expedition. Twenty-two climbers had died, half of them Tibetans or Nepalese acting as porters or in another support capacity. We didn't see the point to another routine ascent with bottled oxygen, or the point of taking advantage of porters who would sacrifice their lives and the well-being of their families for our benefit.

More than one hundred and fifty climbers have now died on Mount Everest, one death for every five climbers who have reached the summit. Why did we survive when so many others died?

There were a dozen occasions when we should have perished. We were inexperienced. We made horrible blunders and miscalculations. We were poorly equipped and cut off from support or rescue. We fell hundreds of feet and bivouacked two nights in the open on the North Face. We set a record for disasters, and still survived. At the time, I attributed our survival to Woody's incredible good luck.

The fall on Everest was to be just the first of my brushes with death. I would later survive a blizzard in Tierra del Fuego and a hideous car crash in Morocco. I lived through reckless experiments with psychedelics and years of mindless wandering among ruthless and depraved people. Over the years I've asked myself, Why did I survive? Was it dumb luck? The search for a scientific explanation to the question drew me into the exploration of a new reality that's just beginning to make sense to me.

There is a famous Everest quote, "Altitude is the great equalizer." To us it was not altitude, but *attitude* that equalized. For forty years we have remained silent while over seven hundred climbers have attacked and claimed conquest over the Mother Goddess. But what a waste! All the resources, energy, and human lives spent to reach and conquer the highest altitude without the slightest gain in attitude, or even a minimum of respect for the mountain. The world is starting to notice; even the most sympathetic observer recognizes the absurdity of the lost lives and resources.

I no longer believe Maurice Wilson was the maddest of Everest climbers. To me, madness is assaulting Everest in million-dollar expeditions with hundreds of climbers, thousands of porters, helicopters, avalanche cannons, and stepladders. Madness is climbers squabbling over who is to be the first to reach the summit. Madness is turning Everest into a high-stakes business, with inexperienced clients paying up to seventy thousand dollars to be dragged to the summit with their personal Sherpas to carry their packs, cook their meals, and perhaps die for them, with no thought or comfort for the families they leave behind.

Despite our lack of experience and limited resources, we were the first climbers to reach the North Col of Everest without using porters or oxygen. In three relays we each packed 120 pounds of supplies over twenty miles of glaciers at altitudes up to twenty thousand feet. I was the first American, and one of the youngest ever, to kick a foothold in the snows of the North Face. We were the first Everest expedition without a casualty. Nonetheless, we are still considered mavericks, in a league with Maurice Wilson.

Woody was the reason I went on the expedition. As a student in his Introduction to Philosophy class at Tufts University, I was captivated by his spirit of adventure. The first day of class he ran up the aisle of the lecture hall in the snowshoes and parka he had used on an expedition to Mount McKinley, vaulted over the lecture table, and landed on the chair standing up. The class stood and applauded; he bowed deeply and grandly.

On another occasion, he had just explained to the class that the biblical story of original sin in the Garden of Eden was a metaphor for *Homo sapiens* acquiring the consciousness of self that separates them from lesser species. A thunderclap pealed out in the atmosphere above the classroom. Without hesitation Woody stood up, raised his arms, and addressed the heavens, "Ah, thank you"—as if thanking God for validating his hypothesis.

He was famous for leaving the classroom, more often than not one on the third floor, during final exams. After a while the students would begin to wonder what had happened to him only to have him open a window from the outside and crawl into the room, having climbed up the stone facade of the building. Woody was confident, cocksure, iconoclastic, full of adventure, and believed anything was possible.

My first personal encounter with Woody occurred during a cross-country race at Tufts. At the time I was naive and twenty, had been to Antarctica twice, and was confused over which authority figures to follow.

A late fall snowstorm was frosting the grass of the football field, cold air seeped into the locker room, and the smell of wintergreen on the runners' legs permeated the air. Although slush splashed onto the fenders of the Amherst College bus as it pulled into the parking lot, the traffic had cleared black ruts along the roads.

The Tufts University cross-country course covered a loop that started at the athletic fields, climbed up to the parking lot of the chemistry building, followed Campus Drive uphill to Barnum Hall, and, after following city streets behind the dormitories, closed the loop at the chemistry building parking lot.

Our coach, Ding Dussault, lumbered into the locker room, beads of sweat and melted snow on his pudgy face. We called him Mr. Jumbo because he wore a bronze elephant pin on his flapping tweed overcoat. Fedora on his head, cigar in his mouth, he brushed and stamped off the snow with great gusto, even though it had mostly melted.

"Now, listen up," he yelled. "You know these guys beat us once, but that doesn't mean we can't beat them. Take one race at a time, and do what you have to do to win. And you know what that is, right?"

"Yes, sir!" we shouted back in unison.

"If you do your best, that's the best that we can hope for." He continued on: "They only got one good runner; that's your job, Roger; stay with him. Now, the rest of you guys stay packed up and don't let them ahead of you. If they do get ahead of you, then get ahead of them. Execute! Execute! All it takes is running, and you guys already know how to run or you wouldn't be here."

He came over to me. "For Chrissake, I thought I told you to get rid of that thing!" He pointed to the black beard I had grown while on an expedition to Antarctica with geology professor Robert L. Nichols.

"I've thought about it. My girlfriend thinks it's kind of distinguished, you know, like Henry David Thoreau, Pierre Curie, or Karl—"

"Who do they run for?" he interrupted. "You look like one of them fuckin' beatniks from Harvard Square."

The "beatnik runner." I liked the concept.

Ding Dussualt had been on my case about the beard for a year. When I had my picture taken for the yearbook, he had flung his fedora in front of the lens just as the shutter went off and yelled at me. "No captain of no team of mine is going to appear in no yearbook with no goddamn beard on him."

The picture of me as captain of the cross-country team in the Tufts yearbook shows me standing in a snowbank in the parking lot of the gym, out of range of the coach's fedora.

After that, I lost both respect for the coach and heart for running cross-country. He was out of touch with reality and made up for it by being pompous, arrogant, and officious.

"Okay, now make the circle." Dussault coaxed us with a circling motion of his arms. He pulled out a cow horn that looked like an elephant tusk and bellowed like a bull elephant ready to trample the kraal wall.

We clasped hands in the middle of the circle and broke out through the field-house door yelling, "GO, JUMBO!"

The dusky fungal smells of fall had been quick-frozen in the earth, and the air was crisp and clean. Heavy snowflakes precipitated from the pewter mist and floated down between the goalposts.

The Amherst team, mostly graduates of private boarding schools, paraded from the field house with an air of calm assurance and performed jumping jacks in their yellow-and-maroon nylon sweats. The Tufts runners, working-class commuters from Lowell, Lawrence, and Lynn, my hometown, milled around and stretched out in gray sweats.

After stripping down to running briefs, the Amherst runners practiced their starts. I shook myself up and down, trying to relax. I lined up with the others at the chalk line barely visible in the snow-covered grass.

Ding Dussault raised the starter's gun and yelled out, "Ready . . . Set . . ." As the blast of the gun echoed off the chemistry building, he dramatically clicked down his stopwatch. The sound of the gun gave me a rush of adrenaline. The other runners felt it too. We jumped off the start line, and pounded up the snow-shrouded knoll. At first we ran in a line, straight across, breast to breast.

My legs felt loose and moved easily despite the cold. Snowflakes melted on my goose-bumped arms and merged with sweat.

As we ran up Campus Way, an Amherst runner hung on my left shoulder two feet back, breathing deep like a steam locomotive. His thigh muscles bulged with the same intensity as the veins on his forehead. He drove his legs with precision. My neck and shoulders ached from the pressure of blood surging through my arteries. Long-distance running is about dealing with pain, and after the first mile it was constant. The other runners had vanished into the storm behind us.

By the time the sandstone shape of P.T. Barnum Hall loomed up in the snow, I was alone with my own world of fantasies. I remembered the stuffed elephant inside the entrance. P.T. Barnum had named it Jumbo. Barnum's fortune went to Tufts, the stuffed elephant part of the deal.

I slowed down, the Amherst runner slowed down; I sped up, he sped up. He was trying to make me think about him. It was working.

We passed the dormitories, ran through the chemistry parking lot, and as we made the first turn at the field house, Ding Dussault, watch in one hand and hat in the other, ran beside me and yelled, "That's it, Roger; stay in there."

I slowed down. A pain growled at the base of my rib cage as we ran through the chemistry building parking lot. I retreated into my mind to avoid the pain. The odor of acetone, chloroform, and toluene, expelled from the fume hoods, set off a cascade of memories. Electron orbitals 1S2, 2S2, 2P2, carbon. In high school, I was so bored that the hands on the wall clock slowed down. I liked college chemistry. Electrons in two forms: waves that stretch to infinity, particles in discrete orbits. An electron jumps to a lower orbit and gives off color. Each element, a spectrum of colored lines, identified in every known atom, in every known star, in every known galaxy in the universe. That's interesting.

The honk of a car horn on my heels brought my attention back to the pavement on Campus Way. I wished I had yelled out, "Hey! I'm running here!" But I was too tired to utter a sound. The cold air burned my lungs. I was beginning to get a cramp. Why? Why did I care about winning this race, enduring the pain? For Coach Dussault? To validate a stuffed elephant? I didn't think so. My abdomen cramped in pain, and I slowed down.

A group of students were scurrying around in front of Barnum Hall, pelting snowballs at a man on the roof.

"You won't take me alive!" he boomed, scooping up gobs of snow and returning fire like a machine gunner. Laughing, he threw his yellow scarf over his shoulder. "You missed!" Fascinated and incredulous, I squinted up at him. The Amherst runner passed me.

The students unleashed a volley of snowballs upward.

"I'll flunk you all!" He warded off the snowballs with his elbows.

A man came to the open dormer window. "Professor Sayre," he yelled. "For God's sake, have some dignity. Get back in here."

"I'm teaching these students a lesson." He laughed and unleashed another volley at the whooping students. I turned onto the driveway under the eaves of P.T. Barnum Hall.

"Yes! Well, all right, I think you're the one who needs to be taught a lesson." The man slammed the window shut, locked it, and disappeared into the darkness of the third-floor offices, stranding Professor Woody Sayre on the roof. The students cheered and yelped. Their snowballs splattered against the shut window. "Hey, come on, Hoar, give me a break. I can't stay up here all night." Sayre pounded on the window. "Oh, shit, will someone call Grounds and Buildings?"

"No way. Call them yourself or freeze to death, Sayre!" mocked a student. The rest cheered.

Sayre beaned the student with a snowball. "Take that, you ingrate . . . Aw, come on, somebody help."

"Forget it, Sayre."

He noticed me jogging under the eaves, his face lit with excitement, and he roared, "Go, Jumbo!"

I shook my head.

"Go, Jumbo, for Christ's sake!" He swept his arm up as if leading the charge of the Light Brigade and pointed at the Amherst runner disappearing around P.T. Barnum Hall.

"Get that fucker from Amherst!" he yelled.

I looked up with doubt on my face, as if to say, "Which fucker?"

Sayre bellowed, "That fucker, get him."

I took off up the hill.

"Go, Jumbo!" Sayre yelled. "Aw, come on, I'll give an *A* to whoever brings me a ladder!"

It was a new race. I forgot about the pain in my side. Time seemed to pass effortlessly as I flew over slush and puddles. A snowball fight. A snowball fight! A professor! Did anyone call for a ladder? I'd dash into the office and open the window for him.

Yells from the finish line. The timekeeper, frozen, watch aloft in a grand gesture of timekeeping. The Amherst runner looked over his shoulder, face red, carotid arteries swollen, sweat pouring down his face.

When Ding Dussault jumped up and down, waving his hat furiously, I slowed down.

A white VW bug rocked on its brakes at the field house, a yellow scarf flashed in the side window, and Professor Woodrow Wilson Sayre leapt out, his blond hair blowing in the storm, his smile as warm as a solar flare. Somehow, in the last ten minutes of the race, he had gotten off the roof, found his car, and driven the short distance to the field house.

"Go, Jumbo! Go, Jumbo!" he yelled. "That's the fucker. That's the one. Look he's beat!" Woody bounded toward the finish line and stood opposite the timekeeper.

The Amherst runner slowed down, holding his side, and his body went into spasms.

"Yes! Yes!" yelled Woody.

Coach Ding Dussault threw his fedora into the air when I crossed the finish line.

Woody Sayre came over and pumped my hand. "Good race. Nice going. What's your name, son? Mine's Woody Sayre."

"Roger Hart. I'm in your philosophy class."

"Oh, good going. Listen, I'm glad you beat those Amherst guys."

"Yeah," I groused, "it's great to be a Jumbo and all that."

"No, not at all." He leaned closer and whispered in my ear, "Do you know what *Jumbo* means in Swahili?"

"No."

"Hello." He winked at me.

" 'Hello?' "

"Hello, just plain *hello.* Barnum never made up that name, but he made millions from it. No, I'm glad you won because that asshole Professor Hoar is an Amherst man."

I smiled, shook his hand. He held mine and patted the top with his other hand.

"Like climbing?" he asked.

"Climbed every mountain in New Hampshire."

"Good, let's get together."

"All right!" I felt ecstatic.

He dropped my hand, waved with a smile, and climbed into the VW. When he closed the door, his scarf dangled out.

"Woody, wait a minute!" I yelled at him.

He rolled down the window. "What's up?"

"Your scarf."

"Oh, that. Thanks. Hey! Your shoes are wet; better change them." He opened the door, reeled in the scarf, and drove off. His headlights cut a swath through the falling snow.

I jogged to the field house, my brain giddy. The locker room was nearly empty. I pulled off my flats and socks. My feet were red from running in the snow.

The next spring, Woody and I began rock climbing on the granite cliffs of Massachusetts and New Hampshire. Returning from a climb the following autumn, he mentioned the idea of going to Mount Everest. I was immediately interested and would have begged him to take me, but was able to hide my excitement. I played it cool, never one of my strengths. The night he called me at my dormitory and invited me, I told him I'd think it over. That night I wandered the Tufts campus with images of Mount Everest before me.

We still had to convince my parents to let me go to Mount Everest. Like most adolescents, I was experiencing difficulties with my parents, rebelling against them and other authority figures of my youth. They seemed out of touch with the reality of my generation. Smitten by the spirit of adventure, I had been to Antarctica, and was determined to go to Mount Everest with Woody.

The setting winter sun exploded red and gold behind the stark oak trees around my father's house. As I pulled into the driveway through a gap in the pink-tinted snowbanks, my father was stomping around the flagpole in galoshes and a plaid carriage coat. He folded the flag, tucked it under his arm, secured the flapping ropes, and turned stiffly toward me.

"Hi, Dad." I opened the car door partway.

"Hello, Roger." He blew on his hands and retraced his steps in the knee-deep snow without looking at me. He wore a blue woolen cap with earflaps covering tufts of scraggy gray hair. The unlatched buckles of his galoshes clicked as he stepped onto the driveway, and with an ice-skating motion, he shuffled across the veneer of snow. "When is this grandson of President Wilson due?" He headed toward the white clapboard house, his back to me.

"Any minute." I pushed the car door open and stood up.

"I hope he's punctual. Are you planning to go with him?" Without waiting for an answer, he pulled open the storm door, pushed the inner door open, and as he kicked his galoshes on the welcome mat, the refrain "There's no place like home" chimed out from the music box attached to the door.

"I don't know," I yelled after him. "We want to talk to you about it." The music box cut off in mid-refrain as he closed the door.

I slumped back into the car's vinyl seat and let out a sigh. Christ, it was going to be harder than I thought. I stared at the last arch of red sun blazing across the snow-covered lawns. I closed my eyes, and an image of the sun, like a green aurora, twisted in my brain. I remembered making snow forts when I was a kid, and gliding off through the nearby woods on skis.

My father was a scientist and had set up an electronics laboratory in our basement. For most of my boyhood, he soldered wires, milled metal fittings, hunched over his ham radio, headphones on like earmuffs, and transmitted staccato messages in Morse code around the world.

I shuffled along the walkway to the house. When I pushed open the inner door, the refrain "There's no place like home" faded into the theme from *Victory at Sea* on the television in the living room. The smell of pipe smoke set the veins in my temples throbbing. I lingered in the front hallway, where snow impressions from my father's boots melted on the braid rug. I watched out the window for Woody's car. Birds carved and painted by my grandfather—scarlet tanagers, eastern bluebirds, and western grosbeaks—perched among my mother's plants on the windowsill.

The pictures hanging over the hall telephone table showed my brother, Stanley, in cap and gown the day he graduated from MIT, my sister, Norma, in her white wedding dress, and my father in his white-and-black Navy captain's uniform. He had made commander on active duty during World War II, and made captain as head of the electronics reserves at the Boston shipyard. My father's captain's hat gleamed white as mothballs on the shelf of the hall closet.

I heard the scrape of the grate and a burst of sparks as my father threw another log on the fire in the living room. The cracking of the fire faded into the roar of TV battleships blasting the Japanese out of Guadalcanal.

A picture of me, taken at age fourteen, showed me in my Boy Scout uniform with my Eagle Scout badge and the colorful symbols of merit badges stitched to a sash—*Botany, Zoology, Astronomy, Geology,* and *Bird Study.*

I walked toward the kitchen. "Mom, hey, Mom, I'm home."

"I'm in the kitchen, Roger." She was an attractive and flirtatious flapper when my father courted her in the White Mountains of New Hampshire. Now she was pleasantly plump, double-chinned, and wore eyeglasses with rhinestones in the corners, her cheeks heavily rouged.

"Oh, dear, I'm just fixin' up a plate of chips and dip for your Professor Sayre." She made the onion dip herself from Lipton Onion Soup Mix. Lipton sponsored the *Arthur Godfrey Show,* which she had listened to on the radio every morning for the first decade of my life.

"Who is this Professor Sayre coming here anyway, your philosophy teacher? Here give me a kiss." She puckered up, took my face in her hands, and kissed me on the lips.

"I had him for philosophy last term." I rubbed the lipstick off with the back of my hand.

"And he wants to take you away to China or India or someplace?"

"To Nepal."

"Nepal, where is that?" She opened a bag of potato chips.

"Just north of India, Mom." I reached for a potato chip.

"Well, I'm sure they're not Christians there, are they?" She arranged the chips around the bowl of dip in the center of a platter.

"There's some, but mostly Hindus, some Buddhists." I was mildly annoyed.

"Reverend Drake says they pray to monkeys, elephants, and cows, and that's why they are all starving to death. The cows eat better than the people."

"They're not starving to death," I replied, exasperated. "They make do with less."

"And they all have greasy skin."

"They put coconut oil on their hair," I replied. "Anyway, we'll be away from all that. Woody wants to go to the Himalayan Mountains."

"That sounds dangerous."

"If I go, I'm just going to base camp. There's no danger at base camp. It will be just like when we camped at White Lake every summer."

"Oh, my, like camping at White Lake?"

Actually I'd had one of the most traumatic experiences of my life while camping at White Lake. When I was too young to swim, I sat in the shallow water near shore, splashed around, and dug harbors. One clear summer day I looked down and saw the slimy brown body of a leech hanging out of my swim trunks. It was worse than a nightmare. Because it had no visible head, its pumping undulations made it look as if it were penetrating swiftly into my testicle. I felt nothing, which made the sight of it all the more startling. Too petrified to move, pull it off, or stand up, I screamed hysterically—the most screaming of my life. Well, maybe not *the* most. My mother with great agility, deftness, and admirable lack of squeamishness plucked the leech, already plump with my blood, from my scrotum and threw it back into the lake. Since then, I've loathed worms of all kinds.

"Well, are you going with him?" my mother asked, then turned and yelled into the living room, "Bob, don't forget we are watching *Lawrence Welk* at nine."

"I don't know. We, Woody and I, want to talk it over with you and Dad," I answered.

• • •

Both my parents were strict moralists, but my mother was the more religious of the two. During high school years, I attended church with her every Sunday morning. We shared the same hymnal and sang the Doxology together:

Praise God, from whom all blessings flow;
Praise Him, all creatures here below;
Praise Him above, ye heavenly host;
Praise Father, Son, and Holy Ghost.

Then, after the Reverend Drake gave his sermon, we would recite the Twenty-third Psalm: ". . . Yea, though I walk through the valley of the shadow of death, I will fear no evil; for thou art with me. . . ."

Home from church, I read the Sunday comics, *Tarzan of the Apes, Mark Trail,* and *The Phantom.* The steam from the New England boiled dinner misted the windows and formed beads that streaked the glass. The windows became alive with rivulets, like worms, that stopped, started, and changed direction.

If we kids didn't live up to Mom's standards of morality, she pulled a wooden "shaming" chair to the middle of the kitchen floor and worked us over.

My sister got it for mountain climbing in Wyoming with a single guy. When she fell off the Grand Teton and broke her back, my mother said it was the just deserts of sin.

My brother once got interrogated in the shaming chair for staying out all night in his junior year at MIT.

I had gotten it just six months ago. The custodian of Fletcher Hall at Tufts had found a condom in a dorm toilet. In the fifties, in Boston, possession of a condom was a serious lapse of morality.

My mother called me into the kitchen, hauled the hard wooden chair into the middle of the glaring linoleum floor, patted it, and commanded me to be seated. "Roger, you come in here and sit down." The shaming chair was worse than the electric chair. Electrocution, that was it, you're dead. The pain of the shaming chair lasted a lifetime.

"Do you know what Dean Schmitt said when he called today?" She faced me squarely, with a pot holder in her hand.

"I haven't the slightest idea." I shuffled my feet and stared at the floor.

"The janitor reported that you tried to flush a contraceptive down the toilet. A contraceptive! A used contraceptive!" And she waggled the pot holder at me.

My girlfriend, Jill, and I didn't really have intercourse. Our first night together, after skiing at Waterville Valley, we zipped our sleeping bags together in the hayloft of a barn, and began kissing. She parted her lips and half closed her eyes. We were afraid to go all the way. But Jill knew what she wanted and how she wanted it.

I began caressing and kissing her beautiful breasts.

"Stop that."

"Why?"

"Because it's driving me ape."

I did it all the more.

We pulled off our ski pants and I climbed on top.

"No, I don't want to play that way."

"How do you want to play?" I asked.

She drew my fingers down her thigh. She was wet as a French kiss and gently slowly wove my fingers through her folds of flesh.

Then she arched her back and rocked on my hand.

"Oh, I feel wonderful," she said.

We gave each other orgasms all night long.

"Where did you learn to do that?" I asked.

"Girl Scout camp. You?" she said.

"Same place. I mean Boy Scout camp."

We had that in common; we loved camping together.

One night we got up our courage, bought condoms, and tried to go all the way, but didn't like it. We went back to our old ways, the ones that worked, flesh against flesh. I threw the unused condom in a dorm toilet. Strange. In the fifties, it seemed as if used condoms stuck around just to cause trouble. It didn't flush down. The custodian fished it out and turned it over to Dean Schmitt, who put the whole dorm on probation until the guilty party confessed. Me! I confessed; the dorm proctors chastised me as though I were a serial rapist. Dean Schmitt called my mother; she called me to the shaming chair.

"I don't want you ever seeing that whore again," she said.

"She's not a whore," I replied.

"I never made love to your father before he married me. Such dirty things in my family. You have this beautiful big house. We paid your way to college, and this is the thanks we get? Don't you dare ever bring her into this house." My mother threw the pot holder and it bounced off my shoulder.

"We thought we might get married," I protested, pissed-off. I ran upstairs to my room.

The doorbell rang. "I'll get it." I ran excitedly to the front hall and pulled open the door. Woody, snowflakes in his hair and his scarf thrown over his shoulder, smiled the best politician's smile.

"All right! Woody! You made it. Are the roads bad?" I asked.

"Roger, how are you?" He pumped my hand as if I were a registered voter and smiled at me in a way that made me feel like the most important person alive.

"Come on in." We walked to the living room.

"Dad, this is Woody Sayre, my philosophy teacher."

My father set his pipe on the stand, stood up, turned off the TV, and shook Woody's hand.

"Would you like a Seagrams Seven and Seven-Up?" It was plain my father had already had a few.

"Sounds good," replied Woody, and my father shuffled off to the bathroom linen closet, where he kept his bottles of Canadian whiskey.

"Mom, this is Woody Sayre," I said, turning to her.

"I hear you are taking my son away from me," she said.

"That's what I'm here to talk about." He smiled broadly.

"Well, it sounds awfully dangerous to me," she replied. "Did you know *Lawrence Welk* is on at nine?"

"Ah, no, I didn't," Woody said with a confused look on his face.

"Well, he is, and we're going to watch him. Do you want to watch him?" asked my mother.

"Ah, no thanks. If Roger goes, it will only be as far as base camp," he said with great charm. "Besides, mountain climbing isn't all that dangerous if you follow the rules and know what you are doing."

"And you know what you are doing?" My father returned from the bathroom with two highball glasses, ice clinking.

"Norman Hansen, the other member of our expedition, and I were successful in our climb of Mount McKinley. Not to mention"—he turned and he winked at me—"Roger and I have climbed every major peak in the Boston area." There was silence.

"Oh, ah, Professor Sayre, did you know Roger made the Dean's List?" my mother asked with a nervous laugh.

"Probably because of the *A* I gave him in philosophy."

"All the kids have Bob's brains." She pushed the dip toward Woody.

"Thanks, Mrs. Hart." He scraped a chip through the dip. "You can be proud of your kids."

"Here you are." My father handed Woody the Seven and Seven. The light of the fire sparkled on the drinking glasses and on my mother's collection of hand-cut Waterford crystal on the mantelpiece.

"Roger doesn't drink. He's the baby of the family," my mother gushed, and turned to me and said, "Sit down."

I sat in a wooden Victorian chair next to the window. My mother carried the tray of dip to my father; he put his arm around her waist flirtatiously. She pushed it off brusquely and said, "Not now, Bob; don't be fresh."

Good sign, I thought to myself; loosen up, folks.

"So you teach philosophy?" My father turned to Woody.

"That's right." Woody took a sip of his drink.

"I always thought philosophy was a waste of time. I'm a scientist myself." My father swirled his ice and drink together. "There's nothing like the thrill of scientific discovery."

"Yes, did you know Bob invented radar or something?" my mother added.

"Really? That's great." Woody smiled graciously. "Pure thought is not without reason."

"Oh, yes, sure," my father said, confused, "but look what science has brought us—radio, television, cars, airplanes, and the atomic bomb."

"Yes," said Woody with carefully measured tone and serious demeanor, "but did we really need to use it?"

"If the Japanese had thought of it first, they would have used it on us." The muscles in my father's neck started to quiver.

"That's a bit of fallacious . . ." Woody started to say.

My mother interrupted, flustered. "Well, where are you taking Roger? Mount Everest or someplace?"

"I haven't decided if I'm going or not," I said

"No, we're not going to Mount Everest." Woody glanced toward me. "We're going to Gyachung Kang, the highest unclimbed peak in the world." Without the money or the political pull to acquire permission to climb Everest, Woody had received sanction to climb a neighboring peak. He intended to follow an unsanctioned route pioneered by Sir Edmund Hillary in 1952—cross into Tibet and approach Everest from the unclimbed north side.

"Oh, sounds dangerous," said my mother.

My father pulled on his pipe. Over the years, the smoke had stained the ceiling above his chair. He stared into the fireplace, sipped his Seven and Seven, started to speak, then fell silent.

I stared out the window at the fall of snowflakes on the forsythia stalks. I remembered the excitement I felt the morning I had spotted my first Maryland yellowthroat in that yard. For the first seven years of my life I awoke every morning to see chickadees and the bird feeder outside my window.

Then I remembered the hurricane that roared through the yard on September 15, 1944. My father was in the South Pacific. My brother and sister at school. I was home alone with my mother at storm's landfall. I remembered wind moaning in the tops of the black oaks, branches screeching against the windowpanes. Cracks like gunshots. Limbs snapping off and bouncing around. The walls pop and buckle; pressure fronts pull at the studs like a wrecking bar. About to blow the house apart. I never want to be in another windstorm.

"Mrs. Hart"—Woody finished his drink and turned up the charm a notch—"I certainly hope you are aware of what a fine son you have here. I saw him run a hellava cross-country race. He beat all the runners on a very good Amherst College team."

"Oh, yes, he brought home some kind of trophy last year," she replied.

The fire started to die out.

"It is quite unusual to meet a lover of nature of his age." Woody reached for a chip.

"Ready for another drink?" My father drained his glass.

"No, thanks." Woody waved his hand over his glass. Then he turned to my mother. "I want to assure you Roger will only go as far as base camp; he'll be perfectly safe there."

"I don't know; I just want him to finish his degree, get married like his brother, Stanley, who just bought a beautiful house in Silver Spring. You should see it."

"Oh, yes, how nice," replied Woody.

"Stanley got his Ph.D. from MIT. Bob's brains, you know."

"Even though Roger will miss spring term, we've worked out that he can still graduate if he takes physical chemistry at Harvard this summer." Woody set his glass on the coffee table.

"I just don't know," said my mother; "it doesn't sound like the abundant life to me."

"Sure it's abundant," I chimed in. "I get to see India, Nepal, the Himalayas; that's a once-in-a-lifetime opportunity. I mean, money isn't everything."

"Jus' d'pends what you mean by *abundant,* doesn't it, Rog?" My father returned with another drink.

"I'd rather he doesn't go," said my mother.

Since the fire was almost out, I got up, pulled back the screen, and picked up a log. My father put down his drink and grabbed the log from me. "Here, let me do that before you burn the goddamn house down." Woody's smile dropped. I stood up, exasperated, and said slowly, "I'm not going to burn the house down."

"That remains to be seen," my father said with a cavalier lilt. "Don't send a boy to do a man's job."

I flushed with anger. "And I can decide for myself."

My mother's face froze. "Oh, oh, dear. It must be time for *Lawrence Welk*. Come, let's watch *Lawrence Welk* together."

"No. It's not time yet." I replied.

Woody stared at the reflections of the fire in his highball glass.

My father rose up from the fireplace shaking. "We have nothing to say, then." He turned on the television. He sat back down and turned to the screen.

Woody stood up and held out his hand. My father glanced up. "You, Professor Sayre, are a crackpot." He turned back to the television and stared resolutely at the screen.

"Don't bother." I stood up. Neither of my parents looked at me. "We'll let ourselves out."

The music box cut off as we closed the door behind us.

"Well, that went well." Woody whistled and shook his head.

"Doesn't matter; they'll get over it," I replied.

"Hey, Rog . . . we're going to Mount Everest." Woody smiled, spun around, arms in the air, and then slapped me on the back. "Except you; you're only going as far as base camp."

• • •

In pictures taken from the space shuttle, Mount Everest appears to have no height at all. Like a chocolate cake covered with frosting seen from above, the steep layers of black rock do not show up. Glaciers surround the mountain like the snow-covered branches of a tree, making a radiating pattern of white limbs that jut out from the mountain.

With the brains of early primates, the four of us had crept down the branch of the East Rongbuk Glacier, crossed the trunk of the main Rongbuk Glacier, and started up the branch of the West Rongbuk Glacier leading to Nepal. We hoped to reach the Nup La, cross the divide, descend the branch of the Ngo Jumbo Glacier to our base camp, and follow the trunk to the safety of the Sherpa village of Khumjung.

It has been forty years now, but I remember our race against the monsoon. With my mind's eye, I see the low ceiling of dark clouds and the veneer of wet snow that blanketed the upper stories of crumbling chocolate rock. Wet snowflakes covering our eyebrows and beards, the four of us, stooped as if pushed down by the weight of the clouds, plodded ropeless in the labyrinth of hidden crevasses at the top of the West Rongbuk Glacier.

On June 13, 1962, we woke to a golden light and searched anxiously for the cache that contained the food and equipment essential to descend the Ngo Jumbo icefall to base camp.

"It's a cache flag!" Hans Peter yelled excitedly as he raced over the pink snow, ignoring the danger of hidden crevasses.

The rest of us followed his footsteps. Norman, digging with his collapsible aluminum shovel, uncovered a pink-nylon marker flag and followed the wand deeper into the snow.

"Pay dirt!" Norman probed a brown duffel bag.

"Even the ravens couldn't find this one." Woody dropped to his knees.

We hauled the duffel bags to the two orange tents and tore them open as if they were presents on Christmas morning: two climbing ropes, three nylon slings, four pitons for rock and ice, and several dozen butane cartridges.

Next came the food, stuff we hadn't tasted for three weeks: cans of butter, freeze-dried steaks, cheese wedges, slabs of bacon, powdered eggs, almonds, raisins, chocolate bars, Swiss jam in tubes, crackers, and cookies.

As we set up a cooking area between the tents, the last mist lifted off the peaks and the snowfields turned brilliant white. We stripped off our windbreakers and down parkas, all the way to the thermal underwear, and bathed in the warm sunlight. The smell of bacon filled the air, and we licked strawberry jam straight from the tubes.

"Tonight, Pemba's fresh chapatis in base camp," I said.

"And in three days, new potatoes at Chombi's," said Hans Peter.

At last our troubles are over, I thought. We had left fixed ropes over the ice cliffs and crevasses below us. On a good day with an empty pack and fixed lines in place, I could descend the two thousand feet of elevation from the Nup La to base camp in under three hours.

After breakfast, we packed up camp, roped together with a new length of shiny gold line, and headed toward the lip of the icefall. We left what seemed like unneeded food and equipment behind. The mist soon socked us in, and we found ourselves trapped between black gashes of crevasses. On a ten-foot jump, Norman failed to reach the far lip and crashed into the crevasse wall. Fortunately he wasn't hurt, but it took us two hours to help him climb up the opposite wall, all three of us pulling him with the rope.

Glaciers are often described as frozen rivers. Where they plunge over precipices, glaciers break apart into a series of steps, like a descending escalator. At speeds of up to four feet a day, the moving steps of ice break apart into gleaming towers balanced several hundred feet above the maze of crevasses.

The Ngo Jumbo icefall is divided into two main one-thousand-foot-high steps by a flat tread and a moraine along the western canyon wall. On the upper step, towers of ice had toppled and shattered into a jumble of ice blocks that had avalanched through the crevasses and cliffs below. The lower step was a frozen turbulence of chaotically arrayed currents and slipstreams of ice.

The majority of Everest deaths have occurred on icefalls. Even with logs to bridge crevasses, ladders to scale cliffs, and armies of Sherpas to cut steps and fix lines, the Khumbu icefall on the South Col route is the most dreaded and dangerous section of the route. When Edmund Hillary and George Lowe first climbed the Ngo Jumbo icefall in 1952 on their way into Tibet, they described it as "incredible." They considered it more difficult than the Khumbu icefall. It had taken us two weeks to make a route through it without ladders or bridges and another week to carry five hundred pounds of food and supplies up it to the Nup La.

It was late afternoon by the time we crossed the crevasse field at the top of the icefall and rappelled down our first set of fixed lines. Woody disappeared over the lip of the crevasse.

We soon heard him yelling, "Goddamn it, I missed!"

"Missed what?" I yelled down to him.

"A crevasse. I tried to swing over to the lip, but missed."

I looked down. A crevasse had opened up along the base of the cliff. It seemed narrow enough to push out and swing to the opposite side.

"Hold on, I'll get you," I yelled down. If I landed on the opposite side of the crevasse, Woody could toss me his rope and I could pull him over. But no dice. It was wider than I imagined. I ended up swinging into the crevasse beside Woody. "Sorry. That didn't go well."

It took us an hour to climb the thirty feet out of the crevasse. Once we were out, Hans Peter lowered the packs one by one, and I hauled them to the lip on the far side of the crevasse. By the time Hans Peter and Norman rappelled down, it was already dark. We were forced to camp.

Occasionally the cloud bank would lift and we could see under it to the location of our base camp, enticingly close. We took turns yodeling and yelling in the hopes that Aila and Pemba, the Sherpas waiting at base camp, would hear us. After dark we signaled with flashlights.

"Why don't they respond?" wondered Woody aloud.

When we set out the next morning, a large crevasse blocked our way. A long piece of our blue rope hung over the edge. The opposite wall of the crevasse had collapsed, devastating our route.

"God! Look at that." I crept forward on my hands and knees, peering over the edge.

"*Scheisse,* always something." Hans Peter crawled up beside me. "Things have changed."

"What happened? Where's the rest of the rope?" asked Woody.

"Look! There, what's that?" Hans Peter pointed to a frayed rope protruding from a jumble of ice blocks. The crevasse had opened, snapped the rope, and then collapsed on top of it.

"Time for a new rappel." Woody peered into the crevasse. "How far?"

"Fifteen feet." Norman stood beside us and studied the ice below. There were no recognizable features with which to scale the distance. The very bottom was blocked from view by an ice ledge that jutted out about ten feet below us.

I moved side to side, like an owl peering into the forest at night. The image of the lip of the crevasse moved in the opposite direction relative to the bottom, whose image barely moved at all. Experience had taught me the farther an object is away the slower the image of it moves. "More like forty," I guessed.

"Whatever." Woody drove a tube piton into the ice, tugged on it, and tied a rappel sling through it. "Who's going?"

I knew that if the details were taken care of, there was no danger to rappelling. With a piton locked in place as an anchor, a metal carabiner holding on to a nylon sling, and the rope looped through it to the bottom of the cliff, it was perfectly safe. I knew that. Yet my sister had broken her back when her rappel sling broke on the Grand Teton in Wyoming. Pitons could pull out, carabiners could snap, belt harnesses could come undone. I went through it all in my mind again, checking the details. Evidently Hans Peter and Norman had similar thoughts, because they were looking thoughtfully at their feet. I've rappelled thousands of times, miles altogether, and each time I've had to screw up my courage. I searched my mind for the right image. It was easier to watch someone else do it first.

"Christ!" Woody shrugged, doubled the rope through the rappel sling, tossed it over, and peered over the edge after it. "It's further than it looks. Hans? Norman? Who's going first?" He offered the rope to each of us in turn. "No, nobody. What the hell, here I go."

It was Woody's favorite situation, facing unspeakable danger with dubious technology. He harnessed to the rope, stepped backward, and lowered his weight onto it.

Recent recruiting advertisements show U.S. Army soldiers rappelling down ropes with great speed and agility. The message? Join the Army and have fun rappelling. Fun? For us it was a dangerous necessity—a staggered hop, a long pause panting, studying the cliff face; then another short hop.

As soon as Woody lowered his full weight onto the rope, the piton anchor slid, slick as a whistle, out of the ice. Until he got over the lip, Woody would be pulling up on the piton.

"Woody! Stop!" I yelled.

"What?" Woody grabbed the rope with both hands and pulled himself back up.

"Piton's pulling out." I stomped on it and held it in place with my boot. "Okay, go ahead."

Woody disappeared over the lip. After a few minutes, he shouted unintelligibly. Hans Peter leaned over and yelled down, "What's that, Woody?"

"What's he saying?" I asked Hans Peter, and stepped forward. The piton slid out again. I stomped it back. Stay in there goddamn it, I thought to myself.

"Roger wants to know what's happening," Hans Peter yelled down. He leaned forward to hear Woody's muffled shouts and then turned to me, saying, "He says he's at the end of the rope. He wants more rope."

"More rope? No more rope." Since he was hanging from the end of the rope, there was no way to tie on another length.

"No more rope!" Hans Peter yelled down.

"He says he could jump, but he can't tell how far it is to the bottom." Hans Peter turned to me. "He says, 'Fuck it,' he's going to jump."

A banshee-like shriek filled the air.

"He jumped." Hans Peter looked down. "It was only a few feet. He's okay. He's lying on the snow with an icicle on him. He's moving. He's holding his head. There's blood coming out of his head. He's waving. He's okay. . . . Oops, he collapsed on the snow."

Norman pulled up the rope, tied the second one to it, and disappeared over the lip of the cliff.

"Norman says Woody's okay," said Hans Peter. "He got hit in the head with an icicle."

Hans Peter lowered the three packs, coiled the rappel rope around his waist, and stepped over the lip of the crevasse. All the time I had kept my foot on the piton. Who's going to hold the piton in for me? The snow was still soft. Was the piton frozen in place yet? The wind picked up, and gray squalls of snow curled up from below as Hans Peter disappeared.

I had to trust the piton. I backed up to the lip in a crouched position to minimize the upward pull as the rope tightened on the piton. I sat back, and lowered my weight onto the rope. I felt as if I were jumping out of an airplane without a parachute. The shadows of clouds racing up from the valley streaked over the snow. I felt dizzy. I could think only of my sister lying all night on a rocky ledge with a broken back waiting for the rescue team. Could anything be worse? *Damn,* go with the fear. I inched backward until I was over the edge. I closed my eyes, pushed off with my feet, and slid ten feet. The piton held!

I looked down. Woody lay slumped in a snowbank holding his head, the snow splattered with blood. Norman squatted next to him; Hans Peter was sorting out the packs.

"What happened?" I asked as I landed in the soft snow.

Woody pointed to a row of impressive ten-foot-long icicles hanging below the ice ledge. The rope had broken off the largest icicle I had ever seen—six inches in diameter weighing over a hundred pounds. It had grazed Woody on the head.

"Geez, that could have killed you," I said.

"Almost did." Woody lay back on the snow.

"How do you feel?" I asked.

"Dizzy. I'll be all right." Woody held his head.

In all my reading of mountaineering expeditions and polar explorations, I had never heard of anybody being injured by an icicle. Yet, it happened to Woody. How strange. On the one hand, we were unlucky, in that so many obstacles arose; on the other hand, we were amazingly lucky to have escaped serious injury.

That afternoon's snow monsoon isolated us in a somber, dreary world. I was worried that the blow to the head might induce a relapse in Woody.

"Perhaps we should camp here," I suggested.

"At this rate, it'll take a week to reach base camp," said Hans Peter.

"Maybe they can hear us now," said Norman.

We yelled in unison, "Pe-e-emba! Help!"

The only reply was the rumble of a cannonade of boulders as they crashed down from the cliffs. Falling rocks had collected in a chute along side the glacier. They were carried forward to the lower step of the icefall and hurled into thin air where the glacier poured over the cliffs.

I felt as though we were trapped in a cruel version of Xeno's paradox. Every day we set out to reach base camp. Every day some unforeseen obstacle cut the distance we could travel in half. Every day we covered only half the remaining distance. At this rate we would never get there.

As we set up camp, I wished we'd brought more food from the cache. We were worse off than ever—out of food again, with little hope of finding our fixed ropes on the bottom half of the icefall. Ever since I was a child, I believed that if I wanted something too much, I wouldn't get it. It was only by not wanting something that I could get what I wanted. Perhaps I was suffering through some strange karma because I had broken the promise to my father. Perhaps out of guilt, a desire to prove I had done nothing wrong, I wanted to get to base camp too much. Okay now, forget base camp. Who cares if we get to base camp? It doesn't matter anymore. My calculations and plans were wrong anyway. Why bother making them?

The next morning we woke to the crashing sounds from the boulder cannonade below. We packed up without eating, barely speaking. We worked our way down to the flat moraine along the rock cliff between the two steps of the icefall and scanned the bottom section for our fixed lines. The landscape had completely changed.

"No sign of them!" said Hans Peter. "It could take three more days."

We took to yelling again. "Aila! Pemba!"

"Where the hell are they?" asked Woody.

From our location, about halfway down the icefall, we could climb out onto rocky ledges above the Ngo Jumbo Glacier and rappel to the base of the icefall in one long but straight shot.

"What about rappelling down the rock wall?" I asked.

Although quicker, once we got to the bottom we had to cross under the cannonade of falling rocks.

"Too dangerous," said Hans Peter. "Falling rock."

"No more dangerous than the ice cliffs," said Woody.

"It'll be quicker, cut down the odds," I said.

"Belay me." Woody started out onto the rock ledges. He drove in pitons in the dusty crumbly rock and put in a fixed rope.

Norman came along last. When he pulled on the fixed piton, it popped out, leaving him holding nothing but rope. This is getting ludicrous, I thought to myself; he could have been killed. We threw Norman another line; he traversed over to the ledge. I set up the belay with several backup pitons.

Woody looped our last rope through the sling and threw it over.

"I guess I'm elected," said Woody.

"Of course." I was amazed that he still had the will to take the lead in these extraordinarily dangerous situations. He was too diminished in energy to be deriving an adrenaline rush. No, he liked tempting fate at a deeper level. Trust your stars, by God. To merely exist is not enough.

He leaned out over the valley on the rope and pushed off. Not surprisingly, he yelled back up in ten minutes informing us that the rope was too short. Fortunately he was able to find a ledge to stand on.

"I'm thirty feet from the bottom. I'll tie off and set up another rappel," he yelled up to us.

After driving in a piton and setting up a second rope, he rappelled to the bottom. Norman rappelled to Woody's previous station, and Hans Peter and I lowered the packs. Norman transferred them from the first to the second rope, lowered them, and Woody untied them at the bottom. The whole process was painfully slow and took most of four hours.

After untying the last pack, Woody scrambled down the ledges at the bottom of the cliff. A mass of boulders ricocheted down the chute like steel balls in a pinball machine.

"Christ!" yelled Woody, and he scrambled out onto the flat ice. After reaching safety, he spun around, and shouted, "It's a shooting gallery!"

We waited for a lull in the rockfall and then moved fast. Norman rappelled down the bottom rope, grabbed his pack, and ran out from the base of the cliffs. He looked back up as I slid down the ropes, picked up my pack, and ran out to join him. Hans Peter stopped at the bottom of the first rope, pulled it through, and dropped it.

Just as Hans Peter landed on the bottom ledges, the cannonade started again. He started to pull the rope through.

"Forget the ropes!" yelled Woody.

Hans Peter grabbed Woody's pack and started toward the flat. Hans Peter looked up at the source of great crashing sounds and, like a deer caught in headlights, froze at the sight of the mass of boulders firing out from the glacier above.

"Run, Hans Peter," I yelled

"Too late. Duck!" yelled Norman.

He ducked behind a ledge and held Woody's pack over his head. The boulders hung in the air like asteroids and then speeded up and crashed around him. A missile the size of his head landed six inches from his face; others bounced around him, then stopped. All was quiet. He peeked out from behind the pack, and then took off running.

"That was too close," he said as he reached our station. "What about the ropes?"

"Forget them. I hope I never have to use another one as long as I live," Woody said, tired and exasperated.

At last we were at the bottom of the icefall, out of danger. A quick walk to base camp and we could gorge ourselves. What could go wrong now? I asked myself. Norman and I moved ahead of Hans Peter and Woody, who, now that he was out of danger, began to relapse. After three days of reduced rations on the icefall, his energy reserves were finally exhausted. He seemed out of energy and out of will.

Norman and I searched among the ice hummocks for the base camp. We yelled for Pemba and Aila. We thought we recognized the spot where base camp had been, but there were no tents. We circled and came to the same spot; it was definitely the right location, but there was nothing there—no tents, no cache, no food whatsoever. We found a plastic bag with a note from Pemba explaining they had given us up for dead and, since they were out of food and it was wet and miserable, they were retreating to Chombi's house in Khumjung.

I couldn't believe it. Damn! They hadn't left us one tidbit of food or canister of fuel. What could be more ironic? I had promised my parents I wouldn't go beyond base camp and now it wasn't even here! We had struggled for two weeks to arrive at a place that no longer existed.

We spent a miserable night. Snow on the walls of the tent melted, formed puddles on the floor, and soaked into our clothes and sleeping bags. A sudden cold snap could entomb us in ice, like Maurice Wilson beneath the North Col. In the morning, we packed out to the side moraine of the Ngo Jumbo Glacier and cached our packs. Woody, back to his routine of falling asleep, lagged behind. Hans Peter was also lethargic. Norman and I felt fine.

"I can't make it," Woody muttered to Norman. "You and Roger go ahead. Leave your packs and send back help."

"We might have to go all the way back to Khumjung," Norman replied. "It could be a day or two."

"It doesn't matter. We'll be okay." Woody shrugged feebly.

Norman and I went ahead. I was delighted to be hiking without a pack, and bounced down the gravel moraine as if walking in reduced gravity. The haze of clouds opened and closed over tantalizing landscapes of rock cliffs. After an hour, we came upon boulders covered with orange and red lichens and tender mats of green moss growing in their shadows. Farther down, carpets of grass draped the rocks and an herbal smell perfumed the fog that curled around the moist ledges. What a wonder even the simplest life was. Every blade of grass seemed a miracle, an eternity of nourishment. A small bird fluttered by and lit up in song. From the mist above we heard distant yak bells and the strange clicks of the herdsman as he talked to his animals. Pathways of sheep and yaks crisscrossed the green pastures, sprinkled with droppings of an intriguing warm fragrance.

Then the most wonderful smell of all tinged the humid air. "Is that smoke I smell?"

"Yes!" Norman held out arms to the heavens and gave a deep sigh, "Saved at last."

We tracked the smell to a herdsman's stone hut with smoke pouring out from under the eaves of the stick-and-slate roof.

"Must be someone inside," I said.

Norman knocked on the door.

A tall, handsome Sherpa chewing a steaming new potato opened the door. I wanted one of those potatoes more than I'd ever wanted anything. But then again, I knew I better not want it too much.

"Sherpa Chombi?" asked Norman, and he pointed back up the valley toward our base camp. I stared at the potato.

"Chombi?" the Sherpa replied. "No Chombi." Shaking his head, he stepped back into the darkness and closed the door.

We stared at the weather-beaten slats. What heartbreak. This was worse than having no food. Totally dejected, we turned and started down the path. How long could this go on? All right, I told myself, stay calm. Sooner or later we would reach food and safety, even if we had to hike all the way to Khumjung.

Then the door opened again and the man stepped out holding a brass bowl of steaming potatoes. He beckoned for us to enter and handed each of us a potato as we crossed the threshold. Ropes, bindings, thongs, and other articles of the herdsman's trade hung in the layer of smoke floating below the low roof.

A woman in embroidered leggings sat straight on the earth next to the hearth and gently pumped air from a bellows onto the smoldering fire. A young boy sat quietly at his mother's side munching a potato, and occasionally feeding twigs from a small pile of heather onto the fire. The Sherpa motioned to a spot near the hearth and placed the bowl of potatoes in front of us. We ravenously downed the sweet, tender potatoes, the best potatoes I've ever eaten. Our hosts watched us with beaming faces and encouraged us to eat more, before whispering and laughing among themselves as the light of the fire danced in their eyes.

After a while the man stood and gestured for the flashlight that I carried in the pocket of my windbreaker. I gave it to him, thinking that he wanted it in payment for the potatoes. He left the hut and closed the door behind him.

The woman and the boy, smiling, watched us finish up the last of the potatoes. We curled up on the earth floor in front of the fire. We were home at last. Not at home with our families, but at home in the family of mankind, gathered around a warm hearth, secure.

The boy, calm and sure of himself, fit right in with the adults. I wished for the same calmness for me and my children, if ever I had any. He wasn't bored and, in contrast to the children I knew in America, didn't demand special attention.

I dozed off and then woke up, scarcely believing I wasn't inside a tent. After several hours of dozing and staring into the fire, we heard voices outside and saw the light of flashlights through the cracks around the door.

"Norman?" shouted Woody hoarsely.

"Oh, fantastic. You guys made it," I said.

The Sherpa entered the hut, dropped my pack on the dirt floor next to me, and handed me my flashlight. Two young Sherpas carried the other packs in and leaned them against the wall.

Woody entered, weaving back and forth. He shook his head and sank to the floor. "I've never been so bone-tired."

Hans Peter sat down next to me. "Thanks for sending the Sherpas to look for us."

"We didn't send them," I said. "They figured it out on their own. Some of them probably had relatives that worked as porters for us on the way in. Anyways, I'm glad they did."

"They went all the way to base camp to retrieve your packs." Hans Peter smiled and waved at the boy.

The boy waved back, and he and his mother burst out laughing.

"Aren't these people wonderful?" said Hans Peter. "I'm going to stay here for a while."

"It could be miserable—all the rain, the monsoon," I said.

"Why don't you stay with me?"

"Well, you know, the leeches. Is it true what they say about the leeches in the rainy season, that they surround a man if he stays in one spot?"

"Yes, it's true," said Hans Peter. "Aren't they amazing, the way they home in on you, climb over bushes, surround a person?"

"Incredible. Oh, yeah. Incredible. I can hardly wait."

He shrugged and poked me in the kneecap. "Hey! Don't stay in one spot."

The next morning four Sherpas showed up to guide us down the valley of the Duhd Kosi to Khumjung. Woody, even though we were in safe hands, or perhaps because of it, didn't improve but stumbled along and fell asleep whenever our party stopped for a rest.

The morning of the third day, we crossed the Dudh Kosi and began the ascent to Khumjung.

We climbed a trail that switchbacked up through rhododendron and oak groves. As we emerged into open terraces of stone walls that disappeared into the mists overhead, a man wearing white cotton leggings bounced down the trail toward us carrying a thermos bottle. I knew it was Chombi before I realized it was him. He stopped, waved his hands over his head, jumped off a boulder, and ran toward us.

Chombi was the headman of Khumjung and, next to Tensing Norgay, who first climbed Everest with Hillary, perhaps the most famous Sherpa in the world. He had been to the United States accompanying the sacred Khumjung Yeti scalp to the Lincoln Park Zoo in Chicago, where it was examined by zoologists.

By chance, we had first met Chombi in Katmandu, and he accompanied us on our three-week approach trek. We got acquainted playing Frisbee with plastic cocoa can lids. I can still remember Chombi and his two companions running across the fields in ornate waistcoats and knee-high felt boots, tripping, falling, and laughing as they chased down the little pieces of spinning plastic.

It was then that Hans Peter first suspected he had joined the wrong expedition. We were not serious mountaineers in the Swiss tradition, or any tradition, for that matter.

Chombi had put us up as guests in his house and showed us his scrapbook. One picture showed him, bug-eyed, in his best ceremonial Tibetan herdsman hat, flanked by Edmund Hillary and Marlin Perkins, then director of the Lincoln Park Zoo, on top of the Empire State Building. Another picture showed him at a high-society fund-raiser in Chicago. The Lester Lanin Orchestra played swing music in the background while Chombi did his best imitation of the yak stomp, lifting his boots high into the air and flinging his yard-long pigtail about. Women in low-cut dresses, diamonds nested in their cleavage, laughed and cheered him on.

Now Chombi threw his arms around us, pounding our backs with the thermos bottle; then he stood back, scrutinized us, pointed at Woody and me, shaking his finger.

"Ah, ho! You two!" he said, and bent over in laughter.

Woody and I looked at each other, bewildered.

"What?" Woody sat down on a boulder and held his head in his hands, his face haggard, his hair matted.

Chombi poured a cup of yak butter tea from the thermos and handed it to Woody.

"It's wonderful," said Chombi, and he broke out laughing again. "Ho, ho ho."

We broke into smiles, but even the effort of smiling hurt. Woody passed the cup to me.

Chombi led us up the pathway to his house. We ducked low through the entrance doorway into the warm smell of the first-floor yak barn. The mother yak pawed the ground in warning, lowered her head, and shook her horns. We climbed up the wooden ladder to the living floor, and flopped down on a bench beneath a low window next to the hearth. Snow spattered on the windowpanes, melted, and ran down the glass. The shelves on the opposite wall gleamed with large brass pots.

Chombi bounded up the ladder shaking his finger at us.

"I know all about you. You've been naughty boys," he chided with a grin that stretched his mustache straight out at the tips.

"What?" Woody asked.

Chombi laughed and slapped his knee. "You know what I mean." He pointed toward Mount Everest.

"You mean you know we went to Everest?"

"To us she is Chomolungma, the Mother Goddess of the World." Chombi nodded his head gleefully. "Yes, it was in the leaves."

Woody stared at Chombi. "I don't believe it."

"There's still more in the leaves." He pushed together the smoldering embers on the open hearth.

He pumped air from his wood-and-leather bellows; the fire sprang upward and then settled back, flickering pleasantly. The smoke rose to the low ceiling and hung in the wooden beams darkened by generations of smoke. The room smelled of yak butter. Flicking his black pigtail outside his collar, Chombi slid into his yak coat, set his fur-lined herdsman's hat on his head, and grabbed the copper teakettle beside the fire. He signaled to me, and I stood up listlessly. Woody started to stand, but Chombi signaled for him to remain seated. A wisp of steam curled up from the interior of the pot. Balancing the hot teakettle in one hand, he climbed down the worn, wooden ladder backward, to the earthen floor of his house. I followed him.

The snowflakes of the monsoon swirled down from the sky and dissolved into black puddles along the dirt paths. Chombi smelled the air and peered past the terraced, stone-walled potato fields that disappeared into the distant snow. Down in the valley, out of sight over the swell of the mountain slope, the Dudh Kosi roared the wrathful sound of milk-white rapids.

Chombi nodded. "The monsoon is here."

We walked toward the orange-lit doorway of a small shrine. Prayer flags clumped with snow hung limp from bamboo poles.

Chombi ducked low, entered the shrine, and I followed him. Inside, candles burned in front of the silk tapestry of Vairocana, deity of the "Chonyid Bardo," floating on clouds. Chombi pulled the wooden door shut, cutting off the cold air. We sat down on a woven prayer rug. The air of the shrine was warm and sweet with the smell of incense. I felt secure, safe, and peaceful.

Chombi plucked a handful of black leaves from a silver box, sprinkled them in a brass cup, and covered them with hot water from the teakettle. Cradling the cup between two brown, weathered palms, he swished the water and leaf mixture together. His dark, bright eyes closed in prayer, and he drank the concoction in one long draft. He partially opened his eyes and squinted hard and long into the gleaming cup.

A smile slithered over his high cheekbones. He placed the cup back on the altar and beamed with childlike glee. "It's wonderful."

"What?" I asked, growing impatient.

Chombi reached for a carefully wrapped cotton package under the altar. The candles beneath a tapestry of the Mother Goddess flickered and cast an orange glow on the white walls. Chombi poked me in the kneecap and carefully unrolled a light brown animal skin with short fine hair.

"Yeti," he said, patting the skin.

"What do you mean, 'yeti'?"

"You know"—he laughed—"the abominable snowman, bigfoot."

He took out his curved knife, sliced off a clump of hair, and handed it to me. "Here, keep this. Do you want to see the scalp?"

"What scalp?" I asked.

"The yeti scalp, the one Sir Hillary and I brought to Chicago."

"Yeah, why not?"

He stood up, lit an incense stick from one of the candles, and set the incense in a holder in front of a large wooden cupboard ornately painted with drifting clouds. Rhododendron petals covered the shelf in front of the cupboard. Chombi picked up a delicate brass bell. Silvery notes cascaded in the shrine, and my body shivered from head to toe.

"Afterwards you must tell me about Chomolungma," he said.

I was puzzled. "I don't know what you are talking about."

He held a finger to his mouth. He set the bell down and opened the cupboard door. A man-sized scalp, with fur the same color as the skin on the floor, hung on the rear wall. I'd never seen anything so weird, too weird to be true.

"This is the scalp the scientists at the Lincoln Park Zoo studied," announced Chombi, beaming proudly.

"What did they say about it?"

Chombi held his belly and giggled. "They said it was a fake, a stretched goat scalp."

"Why did you try to fool them?"

Chombi burst into laughter. "Ho! Ho! They fooled themselves." He shut the cupboard and grew suddenly serious. "Now tell me about Chomolungma!"

"I don't remember."

"Yes, you do; you felt wonderful."

"Wonderful?" I traced the patterns on the prayer rug with my finger.

"Yes. She's a tiger and attacks men."

"We didn't see tigers, just ravens."

"Ho! Ho! No! No! She tricked you, pushed you over the edge."

I shuddered as the memory of the fall came to the foreground of my mental landscape. "Trick? Are you saying the mountain is alive?"

"Everything is alive. The Mother Goddess is very powerful."

"Mount Everest, alive? I don't believe it."

Chombi rose up, fierce, stern, like a wrathful deity gone awry. "You believe it! Remember now!" He picked up the bell, and as he rang it, the cells of my body burst with silver explosions, the energy propelling me back to the night on the North Col. It was like remembering a dream or perhaps a past life. It was happening to me all over again.

"Woody had let the rope hang slack between us," I started. "Before I knew what was happening, I was falling in the night."

Chombi nodded encouragement.

I remembered the horror of my body scraping over steep, sharp ice. I thought at the end the pain would be unbearable. I screamed and screamed. The terror was the umbilical thread that connected me to myself. No pain, mercy please, no pain at the bottom. The thread of time split with a quick snap of a raven's beak.

"When I was sure I would die," I continued, "I separated from myself. Nothing, nothing left to hold on to. I come into warmth and lightness, floating free of time. I feel no pain, yes, wonderful, connected to everything in the universe. I look down. See myself falling through the night. Such a strange wonderful sight—watching myself falling into the night."

Chombi laughed uncontrollably, rolled about on the floor. "That's it. Ho! Ho! Oh, yes. Oh, yes. That's a good one."

I began to laugh too. It was as if there were two different experiences—one part of me was preoccupied with dying, worried about pain, and the other part of me experienced a euphoria I didn't understand.

Chombi sensed my confusion and offered an explanation: "We simultaneously exist on two levels, the lower self and the true self."

"What do you mean?"

"The world you normally see, even time itself, is an illusion projected by the lower self. *In order to experience the higher world, the lower self must be subdued.*"

"The part of me that was afraid was the lower self?" I asked.

"It's not quite that simple, but you could say the lower self masks the true self and the true nature of reality."

"It was the lower self that screamed?"

"Yes, the lower self does not easily give up the world it created. It doesn't let go of the illusion without screaming. But when the lower self does let go, the true self, along with its connections in the Void, is exposed."

"So once I thought I would die, the lower self gave up its power?"

"Yes. You are very lucky; many people spend their lives without experiencing the higher self before they die, especially Westerners. Your life and science are based on materialism and skepticism."

"Why me?"

"Your friend Woody is an exceptional Westerner. He lives in the infinite potential, even though he doesn't know it. You sensed that and chose to follow him to the threshold of the Void. Chomolungma did the rest."

"But why—?"

Chombi interrupted me. "The discovery of the Void, the Sunyata, is the true goal of life on earth."

"I've never heard of it before."

"That is because Westerners have a short attention span, crave sensation. Yet, unconsciously you seek experiences that introduce the higher self. That is why you risk war, seek adventure, confront death."

"Do all mountain climbers . . . Did Hillary find the Sunyata?"

Chombi covered his mouth and giggled. "No, no, no. This is not possible. He is my good friend. He taught me English and gave me this watch, but I doubt if he'll find Sunyata." He pulled back his sleeve to reveal a Rolex on his wrist.

"Why not?"

"Well, I think it is because he knows too much; he knows too much to find the yeti; he knows too much to be tricked."

"Is the yeti real?" I asked.

"Of course it is! Here is its skin, and you have seen the scalp."

I paused to think; silence filled the shrine; a quiet light shown around Chombi.

"Let me get this straight: we are two things at the same time?"

Chombi bent over laughing. "Ohh. Ho! Ho! Ho! Yes. Every thing in the world is two things at the same time."

"This is all rather . . ."

He pointed at me wide-eyed, tongue flailing, the cosmic trickster, and rolled over backward laughing. "It all depends on how you look at it!"

Several mornings later, the mist had lifted to the highest peaks across the valley of the Dudh Kosi. The potato fields had soaked up the monsoon. Families of Sherpas filed down the stone-walled lanes toward a flat clearing at the bottom of the slope.

Norman walked ahead, and I walked with Hans Peter.

"It's wonderful that I can stay here," Hans Peter said.

"You can keep my parka, sleeping bag, and pack." I handed my pack to him.

"Thank you. I'll mail them when I get back to Switzerland." Hans Peter was happy. He was getting to fulfill his dream of spending time with the Sherpas. Woody had left on an earlier helicopter flight and been dropped off at the hospital in Katmandu, where he was treated for a cracked rib. I couldn't get over our amazing luck. After all we'd been through, no one was seriously injured. Perhaps Chombi was right; Chomolungma was alive and watched over us with one hand and pushed us to our limits with the other.

"Don't stand still or the leeches will get you." I shook Hans Peter's hand.

"I like leeches." He took my hand and pulled me into a bear hug. "You take care now."

The sound of the helicopter reverberated between the peaks. The machine descended below the cloud layers, circled out over the Dudh Kosi, and lit down in the clearing. The villagers circled around and draped white offering-scarves of cotton, *kataks,* around Norman and me. Chombi placed a *katak* around me last, embraced me, and looked deep into my eyes.

"Thank you, Chombi; I won't forget you," I said.

"I won't forget you either. You'll never be the same!" He doubled over with laughter, pointing at me. "Now you have no rope!"

I looked back at him, baffled. *What does he mean by that?* I climbed into the chopper and sat beside Norman.

We lifted off slowly, picked up speed and altitude, and shot off like a dragonfly between the peaks. The rotor wash swirled through the open door and tugged at the *katak.* I wrapped it tighter around my neck and held my hands over my ears to block out the deafening noise of the engine. The slopes below sped past. As we gained altitude, the landscape flattened out into a maplike pattern of abstract shapes. The Himalayas were suddenly small and distant, and I missed them already.

I had to return to the States. I was scheduled to attend summer classes, to make up credits, and to enter graduate school at Yale in the fall. But I should have walked back to Katmandu.

CHAPTER THREE

BLINDSIGHT

How is it possible to be two things at once? Chombi's explanation of my experience on Mount Everest didn't make sense to me, a fundamental materialist, the type of person who thought duality was missing the point. I believed in the scientific method of reducing things to their individual parts and trying to understand how they worked together by action and interaction. Not that I thought Chombi was wrong—the idea of the mind separating from the body at the time of death explained my experience of watching myself slide over the icy cliffs of Mount Everest—but it didn't seem relevant to my life. I was trained to be skeptical of unproved hypotheses. There was no known physical explanation for something being two things at once. I dismissed the experience as an anomaly of perception due to unusual conditions: UV radiation, lack of oxygen, cosmic rays, or even a high geo-electric potential. All these had been shown to be most intense at high altitudes.

After returning to America from Nepal, I was embarrassed to talk to others about my out-of-body experience; it seemed too weird. I never even discussed it with Woody, who might well have had a similar experience. Nonetheless, the experience of Everest stayed with me and grew like a sprouting seed in my psyche. Something important had happened. My life had changed; my graduate studies at Yale seemed meaningless. I was especially appalled at the greed and ambition of my fellow graduate students.

Most of them wanted to get their degree and a high-paying job without deference to gaining knowledge or exploring new ideas.

Unexpectedly, two courses at Yale did interest me, Quantum Mechanics taught by Professor Wang and the Statistical Thermodynamics course taught by Professor Lars Onsager. I was taught, contrary to previous instruction I had received, that the great problems of physics had not been solved. In fact, we still do not know what gravity, electricity, matter, energy, and electromagnetism are. We know how they behave, but we do not know what they are. The laws of physics are merely descriptions, and they do not determine events in the universe any more than a basketball commentary determines events in the game. I became convinced that anybody could make a contribution simply by thinking about basic problems.

Physics had shown that matter is mostly empty space. If an atom was proportionally magnified to the size of a barn, the nucleus would be a pea inside the barn and the electrons would be specs of dust orbiting the outside. Most of the atom is empty space. The boundary of the atom is defined by speeding swarms of electrons moving so fast they could circle the earth in a minute. Each electron leaves a trail, like a child writing his name with a glowing stick lit in a campfire.

I learned that light is both waves and particles. Was this an example of something being two things at once?

The wave nature of light was established in the seventeenth century by Sir Isaac Newton, who, using a glass prism, broke light apart into the rainbow-like spectrum of colors. Using a second prism, he recombined the colored light into white light. The wave nature of light allows it to be focused through lenses in telescopes, cameras, microscopes, and the lenses of our eyes.

In 1801, Thomas Young confirmed the wave nature of light in the famous double-slit experiment. In a darkened room he shined a light on a piece of metal with two closely spaced narrow slits. Like ripples spreading out from two pebbles thrown into a pond, light waves passing through the two small slits overlap and interfere. When the crest of one wave adds constructively to the crest of another wave, a bright band appears. When the crest of one wave matches up with the trough of another, they cancel out and produce dark bands. The zebra-stripe interference pattern of dark and bright lines that Young obtained demonstrated that light passed simultaneously through both openings as waves.

Light also exists as particles, photons. In 1905, Albert Einstein published proof of that, in relation to the photoelectric effect. A beam of light shining on a metal energizes the electrons, producing an electrical current—which is similar to how solar voltaic cells operate. If light behaved only as waves, the energy of the electrons in the metal should increase in proportion to the brightness of the light beam: the brighter the light, the more energetic the electrons. But this is not the case; instead, the brighter the light, the greater the number of electrons energized, each with the same energy.

It's as if the electrons are swimmers floating in the quiet water of the electromagnetic field. Light churns the water up into waves and throws the electrons out of the water like swimmers onto a beach. As the intensity of light increases, instead of making one large wave that throws a few electrons out of the water with great energy, it produces lots of identical little waves that throw lots of electrons out of the water, each with the same energy. Einstein proposed that light exists as packets of energy, which he called quanta—we now call them photons—and won the Nobel Prize for his discovery in 1921.

The particle nature of light was confirmed in 1923 by Arthur Compton, who showed that light bounces off atoms the way billiard balls collide with one another, an effect known as Compton scattering.

Light behaves both as waves, ripples of energy that extend through the universe, and as particles, discrete packets of energy that bounce off atoms. How can this be?

The mystery deepened when, in 1926, Prince Louis de Broglie showed that electrons also have a dual personality, behaving as waves as well as particles.

The idea that electrons are particles goes unchallenged. We see their impacts as dots of light on the screen of our TV sets or computer monitors. As an electron beam sweeps the screen, it is rapidly turned off and on to create dark and light shapes that the brain interprets as recognizable objects.

De Broglie concluded that electrons following orbits around the nucleus of the atom are actually in wave forms. Their orbits are positioned so that the crests and troughs of the wave in any given orbit are continuous, matching up without overlapping. Only orbits with circumferences in whole units of the wavelength are allowed. Thus electrons orbit at discrete intervals from the nucleus.

As electrons jump from one orbit to another, they absorb and emit photons in discrete colors, energies. If electrons did not exist as waves, we would not see pure colors.

The wave nature of electrons was confirmed by double-slit experiments that showed that they made zebra-stripe interference patterns when they passed through two small apertures.

Because of their wave nature, electrons can be focused with magnetic lenses, a discovery that eventually led to the development of the electron microscope.

Here was another example of something that is two things at once; an electron is both a wave and a particle. As waves, electrons spread out over space and time. As particles, electrons are located in one place in space and time.

It is now generally accepted that all subatomic particles—not just photons and electrons—exist as both waves and matter. The theoretical description of particles behaving as waves is called quantum mechanics, and is the most successful physics theory ever developed. It has been the basis for startling discoveries.

Normal logic does not apply to the wave nature of matter. Two waves can interfere destructively, canceling each other out. One plus one can add up to zero. Matter can blink out of existence when two wave forms overlap peak to trough. States of matter may exist in the universe that we have not detected because their wave forms are canceled out. On the other hand, by filtering out one of those wave forms, the other could appear as if by magic from nowhere.

Perhaps the most startling success of quantum mechanics was the prediction of a new particle, a type of *quark,* the constituent particle of neutrons and protons in the nucleus of atoms. In 1968 Sheldon Glashow, a physicist at Harvard, surmised that the wave forms of two quarks, the strange quark and the charm quark, were destructively interfering. He predicted that by filtering out the wave form of the strange quark, the wave form of the charm quark would manifest and be detected as a particle. So convinced was he of his hypothesis that he pledged to eat his hat if charm was not detected within two years of his announcement. He did not have to eat his hat, and over six quarks have now been discovered.

The basic question is: When and how does a wave turn into a particle? This is called the collapse of the wave function.

In Copenhagen in the 1920s the physicist Niels Bohr proposed that there is no deep reality, only a vague system of potentials and probabilities that sweep through the universe. This view became known as the "Copenhagen interpretation" and is called observer-created reality.

Albert Einstein, on a purely intuitive basis, disagreed with Bohr in a series of debates in the 1930s. Einstein believed in the existence of a deep reality and announced, "God does not play dice!"

Bohr shot back, "Einstein, stop telling God what to do."

Instinctively, I felt Einstein was right, there is a reality independent of the observer. If the world is an illusion, as Bohr claimed, how can two individuals experience the same phenomenon?

In the 1950s the great mathematician John von Neumann analyzed the wave collapse and proved that under normal circumstances the collapse of the wave function had to take place in the mind of the observer. In 1961, Princeton physicist Eugene Wigner concluded that consciousness collapses the wave function.

If the mind collapses the wave function, then mind must be made of stuff that does not collapse.

How can something be two things at once? Strangely, my graduate studies in quantum mechanics at Yale, rather than closing the issue, opened it wider. Could the dual nature of matter be related to the duality I felt during my fall on Mount Everest? For example, could the body correspond to matter and the mind to waves?

Perhaps our minds, or parts of our mind, survive death as quantum waves. I was sure there is a reality after death; I had experienced thoughts and feelings, awareness. After thinking I was dead, I thought about *being* dead. Perhaps death is the end of wave function collapse. There must be a deep reality, something more than waves of probabilities.

I needed clarification and scheduled a private conference with Professor Wang, my quantum mechanics instructor.

From my boardinghouse near Whalley Avenue in New Haven, I walked along the abandoned railroad tracks that divided the Yale campus from the run-down houses of the New Haven ghetto. I felt strangely off-balance, on the edge of two social levels—the Gothic opulence of Yale and the dire poverty of the ghetto. How could Yale just ignore this social inequality?

The Gibbs Laboratory at the north end of Science Hill, one of the oldest science buildings on campus, reeked of the smell of organic solvents that over the years had seeped into the lab benches and stone floors of the hallways. When I knocked on the door of Professor Wang's office, I noticed that the frosted glass panel was embedded with heavy wire mesh to keep it from shattering during an explosion.

Shy and distracted, Professor Wang, in a brown corduroy jacket and blue silk tie, opened the door.

"Yes, what is it you want?" He glanced back at a string of equations on a dusty chalkboard, obviously annoyed at the interruption.

"I have an appointment for a conference." I stepped into the office.

"An appointment?" He searched through the piles of papers on his desk for his appointment calendar. He threw his hands up in despair when he couldn't find it and swept a lock of shiny black hair off his forehead. "What do you mean 'appointment'?"

"I'm in your quantum mechanics class and I need help understanding the class materials."

"Well, you should talk to the teaching assistant." He faced me, hands on his hips.

"I did and he referred me to you. I spoke to you personally and you scheduled me over a week ago," I replied.

"Okay, okay, what help do you need?" He flopped down, exasperated, behind his desk.

"Wave functions describe the position of particles—photons and electrons, for example. Correct?" I asked.

"Correct, waves of electromagnetic radiation collapse to form particles or photons when we put a detector in the path of the light." He picked up a pencil, tapped it, and glanced over his shoulder at the blackboard.

"Such as our eyes?" I ventured.

"Correct. The nerves in the eye are stimulated by photons, by particles that are interpreted as light by the brain."

"So," I said, "whenever we perceive something, we are looking at light that has a collapsing wave function throughout the universe. Each of us must do that billions of times a day."

"Yes, at least billions of times a day. What of it?"

"When does this collapse take place, what causes the wave function to collapse? That is what I want to know."

"You and every physicist in the world. Nobody knows, even John von Neumann couldn't figure it out. He concluded, as did

Eugene Wigner, that the mind of the observer, thought itself, collapses the wave function. Ridiculous!"

"In other words our minds participate in the creation of reality. This sounds like science fiction; actually, it's stranger than fiction," I said.

"Like Niels Bohr said about quantum mechanics, 'If you don't think it's strange, then you don't understand it,' " replied Professor Wang.

"In other words, nobody understands the collapse of the wave function," I said.

"Correct. We don't understand it because we can't understand it. It is a waste of time to even think about it. The collapse of the wave function must be faster than the speed of light. But Einstein's relativity forbids faster-than-light transmissions. That is why the wave functions are not real, just probabilities."

"Faster-than-light transmission is required for there to be a deep reality, a reality beyond probability. Is that right?" I asked.

"Yes. Now, listen! This class is concerned with wave mechanics, the formulation and prediction of the dynamic attributes of subatomic particles. We do not address the collapse of the wave function. That is best left to philosophers or psychologists. It is not a matter for serious science."

"How do I know I 'can't understand it' unless I try?" I couldn't believe Professor Wang would not want to try to understand the collapse of the wave function. Who wouldn't be curious to find out if the universe is founded on reality or mere probability?

"You are just being foolish. Einstein couldn't understand it," replied Professor Wang. "What chance do you have?"

"Well, at least I'm going to try." The more Professor Wang insisted that I couldn't understand it, the more determined I became to understand it.

"Tell me just one thing. The collapse of the wave function must take place instantaneously. Right?"

"Correct, faster than the speed of light."

"Electrons space themselves around the nucleus of the atom because of the crowding of their wave functions. The information on the spacing of the wave function must be transmitted instantaneously, faster than the speed of light—between all ninety-two electrons in an atom of Uranium, for instance. "

"Yes, faster than the speed of light. That is precisely why wave functions cannot be real. They are just waves of potential or probability." Professor Wang was repeating himself.

"Einstein opposed this idea," I said.

"Correct. But since he himself proved nothing can move faster than the speed of light, he could not develop a satisfactory alternative theory. Listen, the world is divided into the classical realm, our everyday reality, and the quantum realm, represented by fictitious proxy waves and density presences. As physicists, we deal only with what works. You'd better focus on reality."

"Thank you for your time, Professor Wang. "

"You are welcome." He jumped up, erased a term on the blackboard, and scribbled out a new one as if I'd never been there.

It was clear that he wasn't going to help me. I wanted to know if there was an underlying reality to the universe. Who would not be curious about this? The view that there is no reality without the observer, as postulated by Niels Bohr, seemed arrogant, an attempt by self-centered physicists to reposition themselves at the center of the universe. It was egocentric to believe that the world, even God, was created by the observer.

I wandered the streets of New Haven and ended up on Commercial Street near the school of art and architecture. The poured-concrete walls with the imprints of plywood cement forms conveyed the same damp Gothic pretension as the rest of the campus, even though it was modern. In the museum part of the school, the original of Van Gogh's *Night Cafe* hung in place. Like a jewel out of place, it reflected the humanity of Whalley Avenue, not the elitist scholarship of the main campus. In portraying light as swirls of color around lamps, it seemed as if Van Gogh had directly perceived the world of quantum waves, the underlying reality of the universe.

Quantum mechanics was challenging the school of reductive materialism, of which I had been a member since my high school science classes. Reductive materialism maintains that things can best be understood by reducing them to their simplest parts and then reassembling them logically. Quantum mechanics had shown that it is possible to subtract something from nothing and end up with more. Sometimes the "more" is different; incalculable properties emerge when the individual parts are added together.

Scientific observations are supposed to be made with the observer sitting coldly detached in a sterile laboratory, calmly and rationally recording them. This method has provided a great deal of information but only a partial worldview. Quantum mechanics was showing that the observer could no longer be separated from his or her observations. We continuously make choices about what to observe and how we observe it. Any observation simultaneously ignores reams of data that have gone undetected by us. For example, one million neutrinos pass through our brains every second totally undetected. Our brains and nervous systems, limited to electromagnetic signaling, detect only a small fraction of the universe.

What is the true nature of deep reality? Is the mind of God real, as Einstein maintained? Or is our universe founded on nothing but vague possibilities and potentialities? The question consumed me. Do real waves of energy collapse into real matter? This deep mystery presented itself to me in the most materialistic of places, Yale University, made me all the more restless to understand my experience on Mount Everest.

I decided to research the accounts of other climbers on Everest, and over the years, I compiled a history of dual-reality experiences, accounts of other climbers who seemed to be in two places at the same time.

Frank Smythe, the British mountaineer, gave an account of climbing alone without oxygen at 28,100 feet on the North Face of Mount Everest in 1933. He felt that he was accompanied by a strong, helpful, friendly, second person, with whom he tried to share mint cake. Smythe felt the companion would catch him on belay if he fell and its presence completely eliminated any sense of loneliness, but it disappeared once Smythe returned to camp.

Peter Habeler, who on May 8, 1978, along with Reinhold Messner, made the first successful ascent of Everest without artificial oxygen, felt as if he stepped outside of himself on the summit ridge and that another person, attached to him by a rope, was walking in his place. He was no longer aware of Messner. He felt as if he were having a dialogue with a higher being and saw himself crawling up, below himself, and beside himself higher and higher. The feeling of being outside himself was interrupted for only a few minutes, when he felt a cramp in his right hand.

If it happened to Smythe and Habeler, perhaps it had happened to me.

My experience also seemed similar to that of some patients

who were pronounced clinically dead on the operating table and then revived. These patients reported that they could look down on their bodies and recount events that took place while they were flat-lined. At first these stories seemed incredible and were attributed to an expanded imagination or sensitive hearing while under anesthesia. But some patients accurately reported events outside the operating room, such as the actions of loved ones in the waiting room or the details of a label on the top side of a ceiling fan, or the details of a distant object such as a Christmas tree on the window ledge on the next floor down. These reports, if verified, would be convincing evidence that some part of our mind or consciousness exists independent of the body.

Although I didn't know if I had been clinically dead during the fall on Mount Everest and although there were significant differences between my experience and that of those who died on the operating table, I began to refer to my experience as a near-death experience.

The public's perception of what happened was quite different. An article about our Mount Everest expedition finally appeared in *Life* in March of 1963, after being held up at the request of the 1963 "Americans on Everest" expedition. Woody's account of our expedition, *Four Against Everest,* appeared later in the year. The book was well written and fair, but the mountain-climbing establishment denounced our expedition because we were inexperienced and did not have permission. The public turned against us. Speaking engagements were canceled. My professors at Yale declared that the expedition showed that I wasn't serious about my graduate studies, and the State Department withheld my passport. I had not expected this negative reaction, and I felt dejected.

In the fall President Kennedy was killed. Although I was a conservative Republican and a supporter of Barry Goldwater, I felt as if the rug had been pulled out from under me.

I was an alien in my own country. My near-death experience seemed the most significant thing in my life but irrelevant to everyday existence, which, ironically, was seeming more and more meaningless. As it turned out, my experience on Mount Everest was the beginning of a series of paranormal events that took place over the course of fifteen years. I learned something new, usually against my will, from each experience and began to see a pattern.

Eventually I was forced to change my view of reality and what it means to be alive.

My second extraordinary experience, in 1966, was shared by three other mountain climbers, Paul Dix, Peter Bruckhausen, and Jack Miller, on the Darwin Ice Cap in Tierra del Fuego at the southern tip of Chile. Earth's last unexplored lands were in the archipelago surrounding the Strait of Magellan. When Jack Miller, leader of the National Geographic–sponsored expedition, invited me to join him, I jumped at the chance.

I had met Jack in Greenland in 1959, when we worked for the U.S. Weather Bureau supplying stations in the Arctic. After discovering our mutual interest in mountaineering, we began to explore the ice cap and mountains near the Air Force base at Thule. Jack, a compact, rugged adventurer from Spokane, Washington, had climbed the glaciers of the Cascade Mountains and taught me to use crampons and to cut steps with an ice ax.

Jack shared the joy of freedom that I found in mountain climbing. We loved to pick a line of sight and walk for miles without interruption. As the horizon swept over us, it unfolded in an unending parade of fascinating rocks, moraines, glaciers, and tundra vegetation.

At Thule, we didn't think twice when the line of sight to South Mountain happened to cross the main runway of the Strategic Air Command. After all, we had wandered around on the runway when we first landed in the MATS (Military Air Transport System) transport. Jack and I, ropes over our shoulders, ice axes and crampons in hand, set out on a direct line from our barracks. As soon as we stepped onto the tarmac, sirens went off. Long lines of blue lights pulsed up and down the runways as if they were billboards at Times Square. Police cars surrounded us, their blue lights sweeping across nearby hangars. Air Police, hands on pistol holsters, jumped out of the cars, and confiscated our ice axes and crampons, evidently thinking they were potentially lethal weapons. We were arrested, frisked, and taken in for interrogation. When the Air Police called our barracks for an ID, our fellow field-workers thought it a great joke to pretend they had never heard of us. We waited in a cell until the director of our operations could be tracked down in the officer's club. It was a bonding experience for Jack and me.

Six years later, after Jack read about the Sayre Everest expedition in *Four Against Everest,* he invited me to join the Tierra del Fuego expedition.

Jack had done an impressive job of organizing the expedition. He obtained funding and equipment from numerous organizations and corporations: cameras and film from *National Geographic* magazine, waterproof thermal boots and a grant to test them from the U.S. Army, down sleeping bags from Eddie Bauer, an outboard motor from Evinrude, an inflatable rubber boat from Zodiac. He made special arrangements to reinstate my passport, which had been restricted by the State Department because I had crossed the border into Tibet "illegally" with the Sayre Expedition

Jack wrote to the minister of defense of Chile requesting permission to explore the remote lands around the Strait of Magellan. The minister's family name, Carmona, is a common Spanish name, but Jack misspelled it "Caramona," which means "monkey face" in Spanish. The reply came back from the defense ministry denying permission for the expedition, partly because of the accidental insult, but also because the Chileans were inordinately cautious about their frontier lands along the border with their arch rival, Argentina.

"We'll go anyway," declared Jack with a determined arch in his blond eyebrows. We were mountain climbers and explorers, willing to take full responsibility for our own well-being and forsake any expensive rescue by the government of Chile. We were not responsible for the minister of defense's paranoia about Argentina.

We loaded several dozen duffel bags of equipment and food on a rusty Chilean freighter in Brooklyn. I shared my cabin with Paul Dix, a veteran of some of the toughest routes on Mount Rainier and in the Andes. Despite his well-deserved reputation as an accomplished climber, he was quiet, unassuming, and contemplative. We took our meals with the captain and a few other passengers.

We spent the days of sailing the Atlantic coast dividing up the expedition rations in our cabin. We portioned out oatmeal and raisins into plastic bags for breakfast. Instant mashed potato flakes were the main evening meal. Paul devised a spice system to make them slightly more palatable. He mixed cumin, cayenne, and garlic salt for "Mexican" instant mashed potatoes; cloves, cardamom, cayenne, and turmeric for "Indian" curried mash potatoes; basil, oregano, and thyme for "Italian" mashed potatoes; and black pepper, onion salt, and cayenne for "Cajun" mashed potatoes.

"It helps to burn it a bit," Paul said of the last.

For lunches, we packaged up cashews, pecans, almonds, cheese wedges, and semisweet chocolate. Then Paul pulled a box out from under his bed. "Now for the best part." He opened a carton of candy bars in pink wrappers and handed me one.

"What are those?" I had never seen them in stores on the East Coast.

"Almond Rocas. Here, try one." He handed me a finger-sized morsel wrapped in golden foil.

I peeled off the foil and bit into it. "Hmm good, like a Heath bar dipped in crushed almonds. Let's have another."

"Uh-uh, no, we're saving them for the expedition." He slid the carton back under his bed. Suddenly he wheezed and pounded his chest. His ears turned blue. He pulled a plastic inhaler out of his pocket and sucked on it.

"What's that?" I asked.

"Epinephrine," he wheezed.

"What?"

"Asthma, I have asthma." His breathing slowed and deepened and his ears flushed red again.

"How can you climb with asthma?" I asked.

"Oh, it goes away in the mountains. Besides, this inhaler makes me high as a kite." He sucked on it again and his eyes gleamed like Christmas tree bulbs.

For my part, I was in a bad mood. My relations with family, friends, and country had deteriorated. My parents couldn't cope with my adventurous lifestyle. The U.S. government had held up my passport until I signed an affidavit promising that I would never again violate the sovereign rights of any foreign government.

I stayed in the cabin and ventured out only for meals. I read *Relativity: The Special and the General Theory, A Popular Exposition* by Albert Einstein and reflected on his discovery of relativity. Galileo, in the sixteenth century, put forth the basic concept of relativity by observing that when one is riding in an enclosed cabin of a ship, it is impossible to know that the ship is moving without reference to an outside object, at least the sound of water splashing against the hull. Nothing inside the cabin gives any clue to the movement of the ship. Although Galileo was unaware of it at the time, this is true because the human sensory system can detect only *changes* of motion—acceleration, deceleration. After eliminating the visual perception of change, the feeling of air rushing past the skin, and the sounds of motion, the only apparatuses we have for detecting motion are nerve signals generated in the inner ear and the brain's limbic system, which are insensitive to constant motion. We have no way of perceiving the earth's constant motion without reference points in the cosmos.

Einstein expanded Galileo's concept of relativity to include the notion that light travels with us inside the cabin of the ship always at the same speed no matter how fast the ship is traveling relative to the rest of the world. The startling revelation of relativity is that light always travels at the same speed, just as if we are each enclosed in our own cabin of space-time.

This basic fact of nature has led to strange conclusions about space and time. We don't notice any effects at normal speeds, but as we, traveling in our own cabin of light and space, approach the speed of light, time appears to slow down for travels in other cabins. For example, if we signal at ten-second intervals out a porthole by switching off and on a flashlight, the time intervals would appear longer to the other travelers. Due to our motion away from them, the light must traverse increasingly greater distances. But to us inside the cabin the time intervals are always ten seconds, no matter how fast we travel relative to the other observers.

I watched the flimsy curtains over my porthole sway back and forth across the chipped enameled walls of the cabin. It was true that without their motion and the sound of the engine throbbing deep in the hull, I would have had no idea that I was moving.

Paul would try to cheer me up by telling stories about South America. I had been on every other continent, and this one compelled me with a sense of mystery.

"What part do you like best?" I asked.

"Patagonia was awesome, especially around Mount Fitzroy. But I like the Sierra Blanca in Peru best because of the friends I made there."

"Yeah, I heard. Indians who eat guinea pigs."

Paul laughed. "Yeah, some of them do, in the mountains. My friends in Huaraz eat chicken. The amazing thing is how well they cope with so little."

When the freighter entered the Panama Canal, I ventured out of the cabin to watch the heavy metal doors grind open on the Miraflores lock. We floated into the lock and rode slowly down between somber, rusty walls. When the clean vibrant waters of the Pacific flooded into the lock, my spirits lifted. Cold stagnant currents in my consciousness were overwhelmed by the tropical air and waves of salsa music blasting out from villages bordering the canal.

Two days later we docked in Guayaquil, Ecuador, and the spirit of adventure rekindled in me.

"Let's go to town," suggested Paul.

I was ready to explore. "All right."

We threw cameras in our shoulder bags and bounced down the gangplank. When I set foot on the dock, I had the strange sensation that it was moving backward. It was as if my brain had been compensating for the forward motion of the boat but when that stopped, stationary land appeared to move backward. The brain compensates for motion. It fills in and interprets between discrete images to produce continuity. When motion stops, the "program" continues for a short time, things move backward, until the brain adjusts to the new situation.

Piles of garbage lined the urine-smelling streets around the docks. The bus to town smelled of everything from goats to sweat.

"This place smells bad," I said to Paul.

"Don't worry; you'll get used to it."

The streets of Guayaquil were a confusion of people. Vendors sold barbecued chicken, and fresh doughnuts sizzled in vats of hot oil. The sidewalks were laid out with displays of smuggled goods: shoes from Italy, underwear from France, soap, hairbrushes, transistor radios, watches.

A blind woman sat on a box against a streaked and peeling wall and yelled out in a nasal drone, *"Lotería! Lotería!"* Ornate balconies canopied the sidewalks. Gaps between the slats of wooden walls opened into interior rooms. I imagined mustachioed men in undershirts taking a siesta with sultry Latin women.

"Let's get a beer." Paul pointed to a bar with tables on the sidewalk.

We entered and walked up to the bar. A dog sniffed and burrowed like a mole through the paper refuse the clientele had thrown on the floor. The place was so filthy that I didn't want to sit down at one of the tables, let alone drink out of glass.

"Don't worry; it'll be okay." Paul had noticed my reluctance. "Let's sit at an outside table."

A young boy brought us two drafts. They were dark and rich, and after a few sips, I began to feel better. "They let kids in bars here?"

"The families do everything together, even work in bars."

We watched the scene on the sidewalk around us. Everybody jaywalked, in the extreme, side to side, back and forth. Men in dark slacks and bright shirts and women in flowered skirts and colorful blouses ambled about, hung out on the fractured concrete, and spilled onto the street like proud bullfighters indifferent to the cars that rattled by at nose length.

"What do these people do all day?" There didn't seem to be a direction or purpose to the people milling around us.

"Hard to say," replied Paul. "They just get along."

"I mean, do they work for a living?" A few sat at food stalls, but the rest seemed aimless.

"There aren't that many jobs. You should read the *The Children of Sanchez*, by Oscar Lewis, if you really want to know. They pick up odd jobs here and there."

"I don't get it; why aren't they doing something to improve themselves?"

"The lifestyle is not all that easy to get. They accept things as they are, but I don't think they're lazy. Let's go for a walk." Paul left money to cover the bill.

As we strolled along the riverfront, the smell of sweet smoke filled the air. A group of thin young men in loose shirts squatted in a circle rolling craps against the curb. One of them looked up. *"Puta carajo! Gringo, quiere marijuana?"* He held out a joint.

"Gracias, no quiero," replied Paul, who spoke Spanish very well.

I walked on thinking they might mug us. The man sprang up. *"Probe lo. Tome!"* he insisted.

"Well, just to be polite." Paul stopped, took a hit, and handed me the joint. I had never smoked marijuana, but didn't want to antagonize the gang. After taking a deep breath, I coughed and choked, and they smiled at me.

"Bueno!" I said, and took another drag. They all laughed hysterically.

I felt a little dizzy as Paul bid them *"Adios."* We headed for a hill between the docks and the city.

"Cuidado, muchos ladrones!" Many thieves—they warned us, but I didn't need to be warned.

Banana trees and bamboo huts, their roofs thatched with palm fronds, dotted the hillside. A dirt pathway switchbacked from hut to hut to the top of the hill. As we started up, chickens scattered in front of us and pigs squealed. As soon as our presence was known, children in rags came running at us from every direction. Now we're in for it, I thought; pesky brats.

"Let's get out of here." I quickened my pace as the kids swarmed around us. "Now we'll be hassled."

"Don't worry." Paul turned and smiled at the kids. "It'll be okay."

The kids followed us quietly, smiled gently, and tugged at our sleeves. I expected these poverty-stricken kids to whine and beg, but there was a peaceful humanity about them. They asked for us to take their pictures. When I looked at them through the camera lens, their eyes were bright and full of life, far more so than the eyes of American kids, glazed from years of watching TV. The adults who straggled up behind them emanated a kindness that quenched my fears.

"They should hate us, but they don't." I looked at Paul as he pulled a fistful of Almond Rocas out of his pack and handed them out.

"Hey!" I said, "I thought we were saving those for the expedition." I suddenly had an attack of the munchies.

"Oops!" said Paul as the kids ran back to their houses with their prizes. "Forget you saw that."

"Look how bright the kids are," I said. "I thought the poor were supposed to be miserable."

"I don't know. They're not. Perhaps their families spend the time to make them feel loved," said Paul.

I stared at Paul as he spoke. The words seemed to slow down as they left his mouth.

"By the way, why are you talking so slow?" I asked. "Are you stoned?"

"Me? You're the one that's listening too slow." He looked at me, studying my face. "I'm speaking my normal speed."

Approaching the summit, we passed a drainage ditch running with urine and filled with garbage. I stepped right over it without noticing the smell.

From the bare top of the hill, we looked over the vast expanse of gray banana groves and bayous that stretched south to the Peruvian border. Men, boys, and whole families paddled in the quiet fingers of the River Guayas, silhouetted in a galaxy of solar reflections. I felt as if I were Balboa looking at the Pacific for the first time, a whole new world, a peaceful world.

"Oh, wow!" I raised my arms and spun around. "We're here!"

"Yeah. This is great!" Paul looked elated, his blond hair tousled by the breeze from the sea, the light dancing in his blue eyes.

We started down through the huts and came to a small cantina with a rusty metal roof and beer ads hanging on the palm frond porch. When they saw us, two men stood up from one of the wooden tables. They staggered to intercept our path. One of them put his hand on my shoulder. I spun around thinking he was trying steal my camera, and covered it, ready to bolt.

He quickly took his hand off my shoulder and held it up to his face, pretending to take a picture with it.

"He wants to take your picture," said Paul.

Why not, I thought, it's not my camera, and handed him the Nikon lent to us by *National Geographic*. The man took a picture of Paul and me. Then I took a picture of him with his arm around his friend.

"Vaya con Dios," they said as we turned and headed down the hill.

"Con Dios," said Paul, smiling.

"Yeah, *Vaya con Dios,"* I muttered, my first words of Spanish.

We loped down the hill as dusk hit the streets. I was exuberant, the streets seemed to dance under our feet, and we embraced the falling darkness as though we were at home, snug before the hearth.

"Hey, what happened to all the thieves that were supposed to hassle us?" I asked Paul.

"If you stop to think about it, we're the thieves." He smiled mysteriously and broke into a laugh when he saw the confused look on my face.

Sidewalk vendors lit their white gas lamps, and charcoal grills flared up. The smell of tripe and roast corn caught our attention. We bought some barbecued beef heart—it was delicious.

"So all these people do is stand here selling a few morsels all night?" I asked Paul as we strolled the sidewalk next to rows of vendor wagons.

"These people learned centuries ago to take life as it comes and to be happy with what they have, rather than wanting something they don't have."

"I wish I could do that," I said.

"You can, why not?"

"I need to prove myself."

"To who?"

"To everybody who ever said I was worthless—to my father—to myself."

"Take life as it comes and prove yourself at the same time."

"Do you think that's possible?" I asked.

"Sure."

Swifts darted among the overhead wires of lampposts, changing direction wildly to chase down flying insects. They let out a squeal like an electric spark each time they closed on their prey. The streets turned into a party, the music blared, the streetlights danced, and laughter rang out from every street corner. I got it. Life was good as it is; no need to work on it. My WASP blood had been infected with the wrong attitude. Once I realized that, I relaxed. Delightful chaos danced on the swampy streets of Guayaquil and rippled through the incandescent soul of the night. We bounced along as if we were walking on a spongy bog. Everything's natural, I thought to myself; go with the flow.

That day was my introduction to Latin America. In all, I ended up spending four happy years there. I'm a "type A personality," always on the go, striving for this or that. Latin America relaxed me; a contact "*mañana* attitude" infected me.

We returned to our ship and sailed down the west coast of South America to Callao, Antofagasta, and Valparaiso.

We changed boats in Valparaiso, and as we sailed south along the western coast of Chile, the weather got colder and the waters darker, even though we were in the summer months of the Southern Hemisphere. We sailed passed Cape Deseado, and entered the Strait of Magellan. The weather grew worse, the shoreline more mysterious, and shouting winds kicked up sudden and violent snow squalls that hung in the opaque atmosphere and frosted the metal deck. We nearly froze standing on the deck watching fantastic snow-clad volcanoes and weird crags roll in and out of the mists. Pinnacles of ice thrust out of nowhere as if they had been frozen in the act of leaping. Glaciers and waterfalls flowed down the elaborately carved shoreline. We sailed past the mouth of the River of Sardines.

• • •

In the ship's library, I found an English translation of an account of Ferdinand Magellan's visit to the River of Sardines on November 17, 1520. He had stopped, restocked fresh water, caught spawning sardines, and waited for the return of his vessel's companion ship, the *San Antonio*, which was reconnoitering Admiralty Sound to the east.

When the *San Antonio* did not return to the fleet, Magellan's sailors became uneasy and fearful. He called upon his astrologer, Andres de San Martin, to solve the mystery. After casting a careful horoscope and studying the celestial spheres, the astrologer announced there had been a mutiny aboard the *San Antonio*, that the wounded captain was in chains, and that the ship had sailed back to Spain.

One day, when Magellan's crew was ashore smoking, salting, and packing sardines, the lookout sounded the alarm and announced the approach of a large bark canoe paddled by half a dozen naked Fuegians. A spiral of smoke from embers glowing in the bottom of the canoe curled above the heads of three children peering over the side. Evidently the Indians had seen the swarms of terns, frigates, and pelicans skimming the silver waters churned up by the hundreds of thousands of sardines.

The first time the Fuegians saw the Europeans, they stopped and stared at the great hulls and tall masts draped with ropes like spider webs. Magellan's guard gathered up the Fuegians' harpoons and paddles and brought the Fuegians aboard the ship.

The men were absolutely naked; the women wore small sealskin aprons. Their long, coarse hair was bound up by thongs, and their hairless, leathery skin was smeared with fish oil, which smelled bad but protected them from the cold. Magellan noted how sleet had settled on the naked, pendulous breasts of a young woman nursing an infant, and that the mother made no attempt to brush off a film of frozen spray that encased the child's bare back. The Fuegians accepted some cooked fish and tore voraciously at several roast ducks, but it was evident that they preferred their food raw.

After gorging themselves, they fell asleep uncovered and prone on the hard, water-soaked deck, with fine sleet forming a glaze on their bodies. Magellan's sailors lifted the ingeniously made bark canoe up onto the deck and piled the paddles, harpoons, and knives in the forecastle. During the night the

Fuegians soundlessly disappeared with all their belongings, and afterward the sailors started to believe that they were spirits.

On November 21, 1520, Magellan issued a formal request that the officers respond in writing to the idea of continuing on across the southern ocean without the *San Antonio*. The next day, the astrologer approved the continued exploration of the strait for sixty days.

Several days later they passed Cape Deseado into the ocean that Magellan named the Mar Pacifico because it was much calmer than the tide-and-wind-churned waters of the strait from which they had just emerged.

Shortly after entering the Pacific, the crew noticed two new groups of stars, now known as the Magellanic Clouds. Undoubtedly, the star clouds had been in the sky previously but were not observed because of overcast skies. The astrologer was unable to satisfactorily explain the significance of the newly discovered stars. Many sailors thought the Magellanic Clouds were a bad sign and attributed the many later miseries of their voyage to them. Almost half the crew died of scurvy, and the rest were forced to subsist by drinking putrid water and eating leather thongs from the sails. Magellan was killed during an overzealous and badly conceived attempt to Christianize a hostile tribe in the Philippines. Only one ship out of the original five successfully circumnavigated the globe and returned to Spain on September 6, 1522. The *San Antonio*, with mutinous crew in charge, had returned to Spain in April of 1521. I was astounded that credence had been given to the prediction of the astrologer. The story had obviously been written after the fact.

From the beginning, the Fuegians were disdained by the Europeans. Even before encountering them, the Catholic Church had said they were damned. Any race of men not of the seed of Adam was outside the scope of human redemption, and therefore all Fuegians, like all non-Europeans, were lost sinners condemned to eternity in hell—Fuegians all the more so because they were the tribe farthest from Europe. Dante placed his hill of purgatory in the center of the mythical southern continent, and some think the "land of fire"—Tierra del Fuego—got its name from Dante's circles of hell. The Fuegians' paradise was the Europeans' purgatory.

For four hundred years, Europe's cargoes of gold, furs, and

whales passed through the strait in the holds of frigates, galleons, clipper ships, ironclads, steamships, and tankers. The Europeans squabbled among themselves for rights to the straits of passage but were unified in their contempt of the Fuegians. Even Charles Darwin, during the second voyage of HMS *Beagle* in 1838, wrote, "the Fuegians are the most miserable creatures on the face of the earth" and, along with the *Beagle*'s captain, Robert FitzRoy, labeled them cannibals.

Darwin did concede some remarkable adaptations, perhaps the first applied to *Homo sapiens,* when he wrote, "Nature, by making habit omnipotent, and its effects hereditary, has fitted the Fuegian to the climates and the productions of this miserable country."

The Yahgan, one of the races of Fuegian canoe people, were uniquely adapted with short, bandy legs and heavy knee caps for kneeling in canoes. It is unknown if the Yahgans adapted to continuous life in canoes by growing shorter legs with kneepads, or if they took to living in canoes because they were already suited to that lifestyle. Recent excavations in southern Chile show that people may have continuously inhabited the region for more than twenty thousand years. At the time of the European encroachment, the Fuegians were the culture living the furthest south, and the only people capable of surviving in the wild Fuegian fjords.

Now, several days after passing the River of Sardines, our ship arrived in treeless, windswept Punta Arenas. Paul, Jack, and I disembarked with our equipment and transported it to an abandoned sheep ranch on the outskirts of town. I had not seen such a desolate land since Jack and I were in Greenland. The wind kicked up dust devils that faded into the flat, brown landscape.

Peter Bruckhausen, an Argentinean from Buenos Aires born of a German father and American mother, joined us in Punta Arenas. Peter, blue-eyed with black hair, looked like a monk, especially with his balaclava covering his close-cropped hair. He had previous experience in Tierra del Fuego on expeditions with Eric Shipton, the famous English mountaineer,

Jack and Paul set to work engaging a fishing boat to carry us across the Strait of Magellan. Peter and I set out to buy last-minute perishables. Peter, an avid martial artist, unexpectedly but affectionately practiced his kicks on me as we bustled around Punta Arenas. At one point we filled up five aluminum bottles with white gas for our portable cooking stove. I picked up two bottles just as Peter landed a roundhouse kick a half inch from my face.

"Aha! You'd be a dead man now," he exclaimed.

"Not so." I pointed to the canister of gasoline in my hand. "I'd light you up like a Buddhist monk. Boom! Enough of this karate chop-chop shit." I was getting to like Peter.

"Have you tried the crabs?" he asked.

"Crabs?"

"Spider crabs—*centolla.* Sangria. Lunch!" He pointed to a restaurant on a street corner. With window curtains and linen table clothes, it was like a European restaurant, bright and clean compared to the place where Paul and I had eaten in Guayaquil.

The crab was delicious, almost as good as the New England lobsters my mother cooked every Fourth of July.

After we finished our dessert flan, I asked, "What's our route?"

Peter unfolded a rumpled map on the table and pointed to the cluster of islands that makes up the Fuegian Archipelago. The Andes mountain chain runs straight down the coast of Chile. Where it wraps around the tip of South America, it dives below sea level. What would normally be high mountain valleys are flooded with salt water, creating fjords that crisscross between razor-sharp ridges rising straight out of the ocean.

"We'll rent a fishing boat to take us across the strait to about the middle of Dawson Island." Peter pointed to a large island on the south side of the Strait of Magellan.

"Why can't we drive the Zodiac?" I was referring to our inflatable boat.

"The seas are too rough; we'd swamp her. Besides we need to save fuel for the channels." He pointed to the labyrinth of narrow channels at the southernmost tip of South America.

"What's after Dawson?" I took a sip from my sangria, red wine mixed with grapes, apples, and pears.

"We sail up Gabriel Channel, round the west cape at Punta Anxious, and sail up this fjord, Seno Hyatt, to the glaciers at the base of the Darwin Ice Cap. We break out snowshoes and crampons and, how you say it, hoof it up Mount Yahgan here and Mount Darwin there, the highest peaks in Tierra del Fuego." He finished off his glass of sangria, and his eyes gleamed brightly.

I knew Jack was eager to be the first to climb these peaks. I had a different attitude. I was not that interested in climbing a mountain just to be the first to do so. I had started my career as an expeditioneer in the context of the New England transcendentalist naturalists, the quest for knowledge for its own sake plus something unexpected. I never questioned the value of our scientific investigations and clung to the simplistic notion that advancement of knowledge would eventually lead to progress and some benefit for humanity. But the more I had explored the earth, the more my attitude changed. I realized that there were no new frontiers and no unknown resources that would help the progress of civilization. The expedition to Mount Everest had been a great adventure, even if it did not push back the frontiers of knowledge. I could not imagine the first ascents of mountains in Tierra del Fuego as great adventures.

The sangria sent the blood rushing to my head, and once we were back on the flat streets of Punta Arenas, the land seemed to glow with a gentle energy.

"Is Tierra del Fuego like this?" I pointed to the open expanse of leaden sky shrouding low purple hills.

"No. This is really Patagonia," replied Peter. "You'll know it when you're there. Worst weather in the world. Perhaps you've heard?"

"The worst weather?" I stared up at the quiet clouds with sunlight shining through in patches.

"There is no land to block the west winds around Antarctica. They circle round and round the earth." Peter drew energetic circles with his hands. "They pick up energy from the oceans like a croupier spinning a roulette wheel." He waved his arms around, and his blue eyes got wild with intoxication. "Tierra del Fuego is one little finger of land sticking into this global whirlwind. Who knows? We could get blown off the face of the earth!" He held up one finger, clapped his hands, let out an explosion of air, and laughed uproariously. "You'll love it; next best thing to being in an earthquake."

"What are you saying? Land of wind, forget the fire?" I had an almost supernatural fear of the wind. Maybe this trip would be a great adventure after all.

"Right. The fire left with the Yahgans. Tierra del Viento. It's a wonderful place! Surprise after surprise." He slapped me on the back. "It will be a great adventure. Let's go back to the *estancia*."

Several days later we loaded our equipment onto a spider crab fishing boat. I stretched out on the bow as we headed south across the Strait of Magellan. Two stocky, bowlegged fishermen took turns in the wheelhouse and sloshed about in the murky bilgewater nursing the engine and pumps. The holds, though empty, smelled of stale sea water mixed with diesel oil, the same smell as that of creosote piers at Brooklyn, Guayaquil, and Valparaiso.

The engine throbbed steadily in the depths of the hold as the bow punched the choppy swell. Dusk fell, the wind died out, and the fishermen turned on a small yellow light in the wheelhouse. They made the final course adjustment for Dawson Island, and, when they turned off the wheelhouse light, the running light cast a faint red glow on the varnished deck planks. As we passed into the lee of Dawson, an unusual calm dawned on the water, which, smooth as black onyx, reflected every photon. As my eyes adjusted to the darkness, star after star awoke in the firmament. Soon the Milky Way raged overhead and emerged from the black water like a glow-in-the-dark creature of the deep. A black sliver of land on the northern horizon faded away as we slid south into the center of the stars.

The light from each star takes billions of years to reach Earth. Yet according to quantum mechanics, we are connected to each star and the stars connected to each other in an incredible network of waves stretching through space and time. Wouldn't it be great to see the wave function network?

I found the arm of Andromeda with its galaxy and spotted the much brighter clouds of Magellan floating above the southern horizon. Early in the century, Henrietta Leavitt of the Harvard Observatory discovered that a class of supergiant stars in the Magellanic Clouds, known as Cepheid variables, change size in proportion to their brightness. It was possible to determine the distance to any galaxy that contained a Cepheid variable by comparing the apparent brightness, which decreases with distance, with the absolute brightness calculated from the period of size change. It was discovered that the Magellanic Clouds are the closest galaxies outside our Milky Way, and that the most distant galaxies are speeding away from us faster than closer ones. Similar observations led Edwin Hubble to suggest that the universe is expanding.

It has been estimated that more than 40 million stars have exploded in our galaxy, flinging large quantities of matter into interstellar space.

Strange, I thought, the study of the largest objects, astrophysics, has shown that cold, empty space is expanding as the stars move away from each other. On the other hand, the study of the smallest particles, quantum physics, has shown that everything in the universe is connected through a field of wave functions, and both ideas were formulated, proven, and accepted within the past century.

The dual nature of starlight, that it exists as both waves and particles, holds mind-wrenching conclusions for the role of the observer, you or I, as participants in the creation of the universe. John Wheeler, a Princeton physicist, conceived of a thought experiment called the delayed-choice experiment. Theoretically, starlight could be diverted around a gravitational lens, a galaxy size mass of matter, such that, in wave form, it interferes with itself, as it does in the double-slit experiment. Yet in particle form starlight cannot go around both sides of the gravitational lens at the same time. At the moment of observation the wave function collapses, and the photon must choose which side of the gravitational lens to pass on. Yet the light passed the gravitational lens billions of years ago. So our choice of how we look at the universe today might dictate an action that took place billions of years ago. Our interpretation of the past is open to choices we make in the present.

The stars rotated around me as the boat changed course. The vastness of space surrounding Earth is incomprehensible. A thin blanket of colorless atmospheric gases separates us from cold, lonely space. If the sun were the size of an orange, Earth would be the size of a grape seed orbiting twenty meters away; the nearest star would be ten thousand kilometers away. There is that much space. I shuddered at our vulnerability. It seems a miracle that we exist at all, a thin skin of life clinging to one of a billion celestial bodies. In a sidereal flash, the atmosphere could be swept off Earth and the whole population would explode and asphyxiate. Strangely, the concept of space I got on the fall on Mount Everest is different from that conceived by science. Space seemed warm, comfortable, full of light, even though there were no visible objects.

The ancient Ptolemaic view—that we are the center of the universe, and the stars surround us on a celestial sphere, like the dome of a planetarium that protects us from the vastness of time and space—must have given the ancients much comfort. It still helps.

As I floated through the center of the firmament on the Chilean fishing boat, I imagined what it must have been like when the Yahgans sailed the strait, each canoe glowing like an ember, and the headlands blazing with signal fires one after the other, on into the night. But now the somber headlands, pitch black, divided the blazing stars from their own reflections in the sea. The Europeans extinguished the lights as if the very acts of discovery and naming were enough to curse the land.

The European invasion was at its height in the 1880s when sheep farmers on mainland Tierra del Fuego established a bounty of one pound sterling on each Indian head—man, woman, or child. Headhunters pursued the Indians throughout Tierra del Fuego. The government decided to stop the scandals arising from head-hunting by turning the captured Indians over to missions. Troops scoured the country, rounded up the natives, and drove them like cattle to mission stations. Families were mercilessly broken up. A Salesian mission was established on Dawson Island in 1891. Unaccustomed to indoor quarters and disease-bearing clothes provided by the Europeans, the Yahgans fell sick with pneumonia and tuberculosis. Strange foods caused sickness and made them believe they were being poisoned. Liquor and venereal diseases also took their toll. The Dawson Island mission was abandoned in 1924.

Ironically, the only race who could survive in this climate without clothing, the Fuegians, died in part from wearing clothing. They had developed the ability to hibernate, albeit for short periods of time, by rapidly putting on fat, partially because of a small stomach and large intestine, and then living off their fat reserves for weeks at a time. This was a necessary adaptation because the weather changed so fast that they were often trapped for weeks on rocky islands or peninsulas without food. They also had a unique metabolism, such that they sweated profusely in the vicinity of a campfire even while Europeans huddled up close, wrapped in layers of wool and fur.

The black mouth of Dawson Island swallowed the streaming stars, and I dozed off. When I awoke it was morning and we were offshore of the Chilean naval outpost. This was the point of departure. The fishing boat would unload us, our supplies, and the rubber inflatable boat. It was up to us to launch and load the rubber boat. Then we would be on our own.

The chief petty officer in charge of the outpost paced up and down in front of the station house, hand on his pistol holster, waving us off.

The boat nudged the wood-slat-and-oil-drum dock. It buckled—some parts submerging, other parts rising up. One of the fishermen jumped out with the bow line

The chief petty officer took his gun out of his holster and held it over his head.

"Oh, my God," said Peter. "If he finds out I'm from Argentina, he'll shoot us all." The Chileans and Argentineans had fought over remote border lands for centuries.

"Stay low," said Jack, "pretend you're an American."

The fishermen looked at us anxiously and perplexedly. They had to pick up their crab pots at high tide and needed to unload us.

I grabbed the stern line and jumped out, staggered a few steps on the sloshing dock, fastened the rope to the bollard, and pulled a duffel bag of expedition equipment to the dock.

A shot rang out. I jumped back in the boat and looked up to see the chief staggering about, a pistol pointed to the sky. I hauled the duffel bag back into the boat.

Jack pulled me back down. "You lay low too."

The chief shouted enraged oaths at us.

"What's his problem?" I asked.

"He says we can't unload here." Paul fished a bottle of Pisco out of his luggage. "Hold on. I can handle this." Paul jumped out and addressed the chief both meticulously and elaborately. *"Muy amable! Muy agradecido!"* he patronized grandly in Spanish. I was impressed. He shook the chief's hand and causally displayed the bottle of Pisco cradled in his other arm.

"Afuera! Prohibido! No pueden," retorted the chief. He did a double take as his drunken eyes focused on the bottle of Pisco.

"Hombre! Por favor. Charlamas un poco. Su officina donde es?" asked Paul.

"Allá no más. Vámanos, hermano." He pointed toward the station house and draped his arm on Paul's shoulders as if he were a long-lost brother.

Paul yelled back, "Go ahead unload. I'll be back."

We piled our equipment onto the dock, unrolled the Zodiac, and inflated it with a foot pump. The fishermen pointed to the rubber boat and shook their heads in disbelief. I started to feel doubts myself. We were going in a flimsy life raft that could be punctured by reef, snag, or iceberg to a place where heavy-hulled boats normally don't go. There was no chance of communication or rescue. The fishermen must have thought we were crazy. Perhaps we were. The motor of the fishing boat revved impatiently. The fishermen said hasty good-byes and pushed off into the sound.

We launched the Zodiac off the dock, rolled two fifty-gallon fuel drums to the middle of the boat, and piled some duffel bags around them. Paul staggered, dead drunk, onto the dock and waved anxiously after the fishing boat disappearing in the distance. "Hey wait for me!" Then he focused on the dock. "Oh. You guys're still here."

"Hello." Jack waved his hand in front of Paul's glazed eyes. "We're all going in this rubber boat to Cordillera Darwin, remember?"

"Oh, no." Horror dawned on Paul's face. "In that thing? It's too dangerous. Oh, no. Not me." He staggered back and forth on the rocking dock, flopped down on the remaining pile of duffel bags, and sank into a drunken snooze.

We bolted the Evinrude engine to the wooden stern plate and connected the fuel line to a red gas can on the floor of the deck.

"Paul, we're going now. Wake up," yelled Jack.

"Leave me alone," replied Paul, annoyed, "or I'll tell the chief you're all from Argentina. He'll put you up against the wall."

"Where is your friend, the chief?" I asked.

"Argentineans! Argentineans! Argentineans!" yelled Paul.

Jack shook Paul. "Shut up now. Enough fooling around."

"Thanks, I needed that," said Paul.

"Can you stand?" I helped Paul up.

Jack and Peter loaded the last of the duffel bags, and I dumped Paul on top of them and he dozed off again.

"I'm driving." Jack jumped into the stern.

Peter and I positioned ourselves on the gunwales on opposite sides of the bow. Peter pushed off with his foot. Jack pulled on the starter cord. The engine sputtered but wouldn't start.

"Open the gas valve," said Peter.

"It's open," said Jack, and he pulled on the cord. The engine sputtered and died.

"Choke?" offered Peter.

"Oh, yeah." Jack pulled out the choke, pulled on the starter rope again, and the engine gurgled. He put it in gear and opened the throttle; we skimmed over the water of the bay. It felt great to be setting off on another adventure.

The Zodiac rode close to the water, as intimate and unobtrusive as a Yahgan canoe. Small cliffs of wave-washed rock, frosted with layers of mussels, barnacles, and limpets, sloped down into clean salt water teeming with swirls of starfish, urchins, and spider crabs. I dragged my fingers through the cold salt water, amazed at the profusion of life below. I felt as if I were looking through a glass-bottomed boat. Sullen kelp forests swayed in the currents and skimmed beneath the boat, waving and separating to reveal nameless fishes darting among the fronds. Purple kelp crabs clung to the stalks; one of them was regrowing a claw lost in battle. How could anything that lived by scavenging dead fish parts taste so good? I knew they had an incredible sense of smell, capable of detecting amino acids from decaying flesh at the level of one molecule in a billion. Perhaps the best smeller in the world is the hammerhead shark; under the right conditions they can detect and home in on prey up to a kilometer away. The oceans are full of pheromones and hormones—we may know less than two percent of them—that trigger mating behavior, bonding in young, and the attack of predators. In a sense, all the life in the ocean is connected through this chemical detection network. What happens in one part of the ocean affects all others.

This ocean smelled as clean and pure as any I have known. The waters were emerald green with a profusion of phytoplankton. Paul pushed himself up on one elbow, shook his head, and looked around. "What's happening?"

Jack laughed and yelled at him, "Hey, Paul. What did the chief say?"

"The chief?" Paul's brow furrowed in the strain of remembrance. "Oh, yeah, the chief." Paul looked off to the distant hills with a thoughtful expression on his face. "He said, 'I don't give a good goddamn. You can go to hell. This whole goddamn miserable land is hell. Go wherever you want to, and to hell with you.' We'd better get going before he sobers up."

"We're already halfway to Wickham Island," replied Jack, scanning the southern horizon.

"Thank God." Paul collapsed back onto the pile of luggage and wrapped himself in his parka.

As we headed out of the bay, the rubber gunwales cushioned the slap of the mild swell, and the Zodiac bobbed pleasantly. I leaned over to Peter. "Zodiac's nice, probably not as good as a Yahgan canoe."

"The Yahgans knew how to cope," said Peter.

"How's that?" I asked.

"Keep your bow up, and stay naked."

"Must be a tad cold, without clothes," I ventured.

"Aha! Second part of the Yahgan strategy: carry a fire in your boat."

"Good idea."

"No fires in the boat!" barked Jack.

"Okay, we'll wear clothes," I said, "except when posing for *National Geographic.*"

"These parkas will keep us drier than the Yahgans ever dreamed of," said Jack, smiling. He had got a donation of Gore-Tex, the wonder fabric that breathed out perspiration while shedding rain. He stitched the parkas together himself and coated the seams with beeswax.

As we entered the open water of the strait, whitecaps and the spray of salt foam danced over the bow. In the deepening waves the boat rode sluggishly, like an overloaded scow. Jack frowned with grave concentration as he steered the Zodiac around the largest swells. As we crossed Meskem Channel and started across the open water of the Whiteside Channel, the west wind picked up and swells crowned with plumes of foam broke over the bow.

Suddenly a squall hauled up in front of the sun and took a bead on us right down the strait. A shiver went through me as the sky turned dark and threatening. The wind brewed up choppy water boiling with swells. A wave crashed into the boat and puddled among the fuel drums. We scrambled to get the cameras and film off the deck.

"The sleeping bags!" yelled Jack.

I pulled them off the sopping floor and lashed them to the top of the luggage pile. Peter cinched the load to the gunwales. Paul jumped up and helped me bail furiously. Wave after wave crashed in on the growing swell, and we soon had a foot of water in the bottom of the boat.

"We're riding too low. Too much weight; we're overloaded!" yelled Peter. "Steer into the waves; go with them." The crashing spray drowned his voice. The boat rocked and twisted in wild contortions and sloshed like a waterbed as it bucked the now mountainous sea capped with foam. We stared anxiously into the foggy dusk, searching for a promise of land. We saw nothing but towering waves on all sides.

Jack, water streaming down his face, looked confused as he tugged back and forth on the tiller. Clothes, cameras, food, sleeping bags, were getting deluged.

"Let's jettison gasoline!" I yelled.

"No, we'll make it!" Jack yelled back.

A huge wave plowed into the boat, and we took the full brunt on the gunwale. The spray splashed over Jack and ran off his blue parka. "See, nothing can sink us." Jack wiped the water off his face and shook his fist at the waves.

"Yeah, but they can swamp us," I yelled back, irritated at Jack's cavalier attitude, "At least slow down." I couldn't get the image out of my brain that if the bow got too low, it could submarine under a wave and tip over.

"Don't think about it and it won't happen," Jack yelled.

"No! Thinking about it stops it from happening," I yelled.

"Screw you, waves!" yelled Jack. "We can't be far from the lee of Wickham!"

The setting sun slid below the clouds, the swells danced in a frenzy of red fire, and the sun sizzled into the water and went out. The humid darkness dampened our souls. The wind picked up, and I felt colder and more forlorn than I could ever remember. We desperately clutched the gunwales as the wild water, roaring and heaving, tossed us about, helpless.

Mercifully, the black shapes of shore pines and ragged ridges loomed up in the haze, and the wind abated in their lee. The swells ceased crashing, and we heard the rip currents gurgling in the cobbles like a marimba.

"We made it!" Jack shook his fist back at the dark strait. The rest of us jumped out without saying a word, pulled the boat up, and tied it to a gnarled pine. The only thing worse than getting our sleeping bags wet would be for the Zodiac to break loose on the tide.

With flashlights we found a camp spot in a grove of beeches, unloaded the boat, and pitched our tent. Our clothes and sleeping bags were wet, but the night was warm under the low comforter of clouds and calm lee of land.

Once we were settled into sleeping bags, Paul fired up the brass Primus stove. The sight of the flame cheered and warmed us. "How would you like your mashed potatoes tonight?" Paul asked cheerfully.

"Whatever," said Jack.

"Ah. How about Argentinean?" asked Peter.

"Okay, Italian. That's close enough. I'll throw in some meat by-product."

After we finished our meal and slid down into our sleeping bags, Jack sat up again and said, "Goddamn it, there's a bump under me. Paul, change places with me."

Paul and Jack climbed out of their bags, exchanged places, and settled down again. Peter and I sat up to watch them.

"Goddamn it," said Jack. "This is worse than before; Paul, change back with me."

Peter started to say something, then stopped. As soon as they were rearranged, Peter broke in. "I think we need to make some changes."

"How's that?" asked Jack.

"You're not driving the boat the right way."

"I'm not?" replied Jack.

"Isn't it obvious?" I said.

"You are driving smack into the waves," said Peter. "You have to slow down until the wave is under you, then speed up and surf down the other side."

"Perhaps Peter should drive," suggested Paul.

"Is it mutiny, then?" asked Jack.

"No, it's not mutiny," I said. "Peter just has more experience than you."

"Well, take some pictures of me driving for *National Geographic,*" Jack said slowly and deliberately.

"With or without clothes?" I couldn't stop myself.

"Don't be—"

"Sorry," I interrupted.

"Okay. Peter is more experienced," said Jack.

We all slid into our sleeping bags and soon dozed off.

I remember I dreamed of Yahgan women, pendulous breasts

covered with sleet, dancing around their campfire dressed in nothing but diadems of heron plumes and white goose down. I woke with an erection.

"Time to regroup." Jack was up at the crack of dawn. "Okay, men, grab your socks and let go of your cocks."

Easy for you to say, I thought.

The others crawled out of their bags, deflated their air mattresses, and started to roll them up.

The fantasy of the Yahgan women lingered in my brain. I could imagine what a ridiculous profile I would present in my thermal underwear. In all the accounts of the great expeditioneers—Robert Falcon Scott, Ernest Shackleton—or the great naturalists—Henry David Thoreau or John Muir—they never mentioned waking up with erections. Were they all saints who never fantasized? Did they have inhuman willpower? Or did they ignore the problem out of politeness and protocol? Think of something else, damn it—socks. Right, find socks.

"Com'on, Roger; let's get crackin'," commanded Jack.

"Give me a break; I can't find my socks." I fished around the bottom of my sleeping bag. I was beginning to resent Jack's officious attitude.

Paul served up breakfast of oatmeal and tea.

The last to crawl out of the tent, I stared up at the clear sky from the door. I stood up and looked out over the sound, where we had almost swamped the night before. Brilliant sunlight sparkled on the azure waters lapping at the satin driftwood along the shore. This was paradise. Parrots fed on the seeds of the winter's bark tree, and exotic hummingbirds skimmed among the flowers. As if in awe of the stunning beauty of the day, the wind had retreated.

"First," Jack commanded, "lay out everything in the sun." We stripped down to our skivvies and soaked up the sun. We draped clothes and sleeping bags over the warm rocks, emptied out the food duffel bags and stacked the plastic bags of food on the sand.

Jack pulled out an air photo of the fjords and straits and studied them, patting his blond hair across his bald spot.

"Paul and Roger," Jack said, "take the fuel drums in the Zodiac and make a cache about fifty miles to the south and get back by nightfall. I'll stay here and work on my journal. Peter start drying things out."

"Sounds good." I liked the idea of seeing new land for the first time, of being the pioneer.

As the Zodiac skimmed along the shore with Paul at the helm, I sat in the bow and watched for submerged snags, shoals, and rocks.

Thick vegetation draped the shoreline. Shore pines shaped by wind and salt spray clung without perch to vertical cliffs. Scene after scene of majestic rocks and lush vegetation rolled past us like Japanese pen-and-ink drawings.

Sea lions barked and clapped from a rock rookery. Paul swung the Zodiac in close. The sea lions barked louder. *"Araruph, araruph, ruph, ruph."*

We barked back at them. *"Araruph, araruph, ruph, ruph."*

A huge alpha male, black-faced, with rolls of orange fur on his neck, craned his head back, bared his teeth, and rocked from flipper to flipper. The females and pups humped to the water's edge and slid into the surf.

"All right, we're going. Sorry!" I yelled at the alpha male. A great flock of cormorants rose into the air with shrieks and the noise of flapping wings. Out of sheer exuberance, I let out a bunch of yelps and yodels. Paul laughed, and yodeled too.

We felt a bump on the bottom of the boat.

"What was that?" Paul asked, cutting back the engine.

"I don't know; I didn't see anything." Then another bump and I heard an eerie gasp for air. *"Awooosh!"*

I leaned over in time to see a shadowy object disappear into the depths. Then we felt another bump, and a huge marine creature rolled, flashed its glowing white belly, and sped out from under the bow.

"Dolphins!" I yelled to Paul.

"They like our yodeling," Paul said.

"Who can blame them?" We felt another stronger bump.

"Their fins; they'll puncture the boat!" said Paul

"No, they roll belly up."

A hundred yards off the bow, a dolphin breached into the air, flipped over, and reentered the water without a splash. We shrieked, yelped, and yodeled all the more.

"It's a race!" Paul yelled, and he gunned the throttle full tilt. An excited burst of exhaust fumes kicked up in bubbles from the stern, and the Zodiac peeled out over the glass-smooth water.

"An excellent sign. Dolphins are very lucky." Paul bounced up and down on the rubber gunwale.

Then we felt five bumps one after the other, five white flashes of dolphin belly crossed under the bow, and in perfect timing, five dolphins leaped into the air, flipped over, splashed down, and disappeared in one synchronous movement.

"Wow! Did you see that?" I yelled

"Synchronized swimming!" laughed Paul. "Give them the gold."

"How do they do that?" I asked.

"They talk."

"Talk underwater?"

"You know, sonar," said Paul.

Then we felt five more bumps and the show started over. The dolphins escorted us the length of Wickham Island, ambassadors of the last true wilderness.

The dusky dolphin, *Lagenorhynchus obscurus,* common in the cold temperate waters of the southern ocean, normally travels in groups of six to fifteen but joins together into groups of three hundred or so to collectively trap sardines. In controlled experiments dolphins have exchanged information without visual clues, suggesting they have intelligence and can communicate with each other.

The masters of sound, fin whales, are known to communicate over distances of five kilometers.

Blind river dolphins, which live in the muddy rivers of Asia and South America, communicate by sound alone. What are the dreams of a blind river dolphin? What images do the careening sounds produce in their brains? How do they formulate a worldview without visual input? Some blind humans suffer from damage to the primary visual cortex. All other components of the visual system are intact. They are convinced they cannot see because visual data does not reach consciousness. Yet when asked to point in the direction of a light source, they are successful above chance levels. Somehow the visual information is recorded in their unconsciousness without reaching conscious levels. The phenomenon is referred to as blindsight.

"That's where we're headed." Paul pointed to a bank of clouds in the southwest. We turned in to a quiet harbor, pulled the Zodiac up on the beach, and unloaded the fuel drums. We watched several dozen plovers work the waves on a sandy strip of the beach.

"Fishing for sandworms," speculated Paul. Like the dolphins, the plovers synchronized their actions. They'd all run down the strand together, tiny feet churning, like battery-powered toys; then they'd break rank and peck away at holes in the sand. As the next wave rolled in, the plovers would turn en masse and run back up the beach, as if afraid to get their feet wet. Then for some unknown reason they would take flight, their wings beating as fast and furious as their legs. The flock would spontaneously change direction in mid-flight, glide like a single parafoil wing to a new spot, and start wave darting again. I wondered if they might be responding to some hidden source of information.

We pushed the Zodiac back into the waves as dusk fell and turned north again. Back at camp we analyzed our position. We had learned our lesson from the heavy swell of the Whiteside Channel. We had to lighten our load. Even so, it was obvious the boat would be in danger rounding Punta Anxious in the open water of the Magdalena Channel. It was decided that Paul and I would backpack over the peninsula while Jack and Peter took the boat around the cape.

CHAPTER FOUR

EYE OF THE BLIZZARD

Paul and I disembarked with our packs on a rocky beach, waved good-bye to Jack and Peter as the Zodiac pulled away, and started up a streambed toward the top of the narrow peninsula separating us from Seno Hyatt. We scaled wave-washed ledges along the shore populated with mussels, barnacles, and sea anemones. Shore pines rooted in the cracks of the cliffs draped over the clean granite. Rocks, darkened above the wave zone by black lichens, led into a forest thickened with chest-high ferns. We pushed our way through the understory. Mosses, ferns, and orchids dripped from every cliff face, and dangled in profusion from every tree branch. We emerged onto the top of a cliff and caught a view of the sparkling water below framed by the silhouettes of pine trees.

As we climbed higher, the slope leveled and the rain forest opened up into peat bogs and meadows laid out in harmony as if planted, pruned, and trimmed by an army of Japanese gardeners. We could see the forest below us, the channel beyond, and past Dawson Island to the Strait of Magellan.

At the ridge crest we emerged onto a windswept meadow of grass. Now I could see beyond the peninsula to where we were going. Below us Seno Hyatt followed a crustal fracture into the heart of the Darwin Ice Cap. We were standing on one of many ridges that radiated like fingers from the white dome of snow. Blue water filled the glacially scoured valleys and led up to massive icefalls, where the ice had broken off as massive icebergs. The summits of Mount Darwin and Mount Yahgan protruded from the ice cap.

The skirts of the mountains were draped with purple-and-green rain forests like the one we had just ascended through. The ridge on which we stood continued westward in a series of serrated peaks that formed the west cape, around which Peter and Jack were driving the Zodiac.

Sparkles of sun from the fjords, the murmur of trees in the forests below, the rustle of the grass, and the drift of clouds excited me. Every one of my twelve million optical nerves exalted in the feast presented by this view. What would it be like to live here?

Harsh wind, steady rain, and rugged mountains guard Tierra del Fuego. No beach is wide enough for a resort or town, no harbor deep enough for oil tankers, and no rock broad enough to give perch for a strip mine.

We had to reach the shore of the sound by nightfall. A squall line marched over the ocean to the west, where Jack and Peter had gone in the Zodiac. The storm darkened and churned the water before it.

We hastened down the side of the peninsula and arrived on the rocky shores of Seno Hyatt before dark. The squall line blanketed Seno Hyatt. We set up camp and searched the black waters of the sound under the ceiling of mournful clouds and anxiously listened for the sound of the Evinrude.

At sunset, a pale beam of sunlight burned a hole in the mists of the sound, parted the drifting clouds, and gilded the chalk-white crests with supernatural beauty. We listened for the stutter of the engine. Dark vapors lifted off the water, swirled, and sealed the rent in the heavens for the night.

We soon heard the Evinrude above the rise of wind, and the Zodiac pulled into sight around a headland. The rough water was behind us.

When I awoke in the morning, I remembered I had dreamed I was a priest of the Inca Empire standing on top of the fortress of Sacsahuaman in the capital of Cuzco. Dressed in high-top thong sandals, ornate cape, and feathered headdress, I looked north toward Panama with a feeling that a dark cloud coming down the coast would humiliate the powers of nature. The scene shifted to a house in Cuzco, where, as a gray-haired man, enslaved by the Spanish conquistadors and employed by the Jesuits, I labored at the work of translating the Bible into the Incan language. I woke up from the dream in terror and drenched in sweat.

After breaking camp, we readied the Zodiac and then pushed east, into the depths of Seno Hyatt. As we skimmed over the water, the peaks closed around us, rising straight up from the sea, as steep and tall as any of the Andes or Himalayas. The change of ecological zones was swift and spectacular. A luxuriant rain forest carpeted the rock walls. Farther up, weather-stunted alpine vegetation formed open vistas. The peaks and ridges were crowned with barren rock and, even in the southern summer, snowcaps.

In the polar regions of the Northern Hemisphere, where the land mass warms up in summer, the glaciers have melted back up the mountainsides. In Tierra del Fuego, cold fingers of the ocean have pulled the ice into the sea. Great thicknesses of snow, from incessant blizzards on the ice caps, pushed glaciers into the black waters of the fjords. The great frozen Niagaras spalled off, leaving ice cliffs of astounding blue cobalt. Thick mists, swirling on the seracs, blocked out the sun. In the dim light, the blue ice glowed from within, as if the waves of the fjord, tearing off chunks of ice, released long-trapped Pleistocene starlight. The sight of blue ice in green forests was startling.

Ice blocks of all sizes floated on the black water. Jack, watching for submerged ice, stood on the bow like George Washington crossing the Delaware.

"How thick is this rubber, anyway?" I probed the rubber of the gunwale with my finger and pushed off a chunk of ice with an oar.

"Too thick for this ice!" said Jack.

Peter cut the engine back to a barely noticeable idle. The salty currents had gnawed away the bellies of some icebergs, shifted their centers of gravity, and rolled them like dolphins. A larger chunk of ice got past us and screeched against the rubber gunwales. Peter cut the engine entirely, the thick fog shrouded any sound, and a heavy silence fell over us. I handed Peter my oar, and he sculled the Zodiac through the icebergs. The only sound was the gurgle of the water spiraling in miniature maelstroms on either side of the oar. Veils of mist shifted, revealing mysterious crags, slices of black water, and steep cliffs of slate that appeared furtively and then silently swept by us.

"Ice on the starboard," said Jack.

"Got it." I fended off a chunk with my ice ax.

The breathless silence invoked a stillness so intense that we seemed to float on the cusp of time waiting for the unveiling of a great mystery. Suddenly a report like thunder boomed through the fjord and shook the mountain walls. The sound of cormorants flapping their wings in the fog came from all sides. Sea lions barked in the distance. Steamer ducks, paddling wildly with their wings and feet, streaked through the black water, trailing silver wakes behind them. An iceberg had broken off and crashed into the water, creating a tsunami of ice. We peered into the thick mist and heard the roar of water and the grinding of ice blocks. Lifting the curtain of fog, the wave of ice blocks sped under, picked the Zodiac up, and spun us around. As we fought feverishly to save the hull, the rubber boat splashed down into the thick slush of ground-up ice. Propelled backward, we grounded just ten feet from the shore we were seeking.

"Wow," said Peter. "That was fun."

"See! We are invincible!" exalted Jack. "No more fun. Unload!"

"All work and no play makes Jack a dull boy," Peter whispered, and winked at me.

We pulled the Zodiac up on a narrow black beach next to a glacier leading up to the Darwin Ice Cap. Where the glacier spilled over the steep rock wall of the fjord, it had broken apart into a jumble of crevasses. After making camp, we scouted a route around the icefall on a rock wall crisscrossed with grassy ledges.

The next day we loaded the tent, a collapsible snow shovel, air mattresses, ice axes, crampons, snowshoes, down clothes, sleeping bags, and enough food and fuel for ten days into our rucksacks, and packed up the ledges above camp to the top of the rock cliffs. Emerging on top, we looked straight down onto the icefall and the waters of the fjord. Standing close to the edge, I taunted myself with fantasies of falling uncontrollably, drove myself dizzy by looking straight down through my feet. My worst fear was that I might give in to an irrational unconscious urge to throw myself off, just to test fate. A surging breeze rocked my equilibrium, and I stepped quickly back to safety.

We followed a rocky ridge upward until it disappeared under a mantle of snow and ice. We found an easy snow slope to the surface of the glacier. Having circumvented all the crevasses, we strapped on snowshoes, trudged upward through soft snow, and watched the peak of Mount Darwin breach the ice cap, its brow furrowed with crevasses.

"At last," said Jack. "Tomorrow we'll have our prize."

We set up camp well out onto the ice cap. While Peter and Paul reconnoitered peaks near camp, Jack and I set out at sunrise, slogging through soft snow. Hour after hour, we plodded to reach the base of the summit cone. Every step we took, we stepped where no one had stepped before. We found our way up the brow, through a crevasse field, onto the summit ridge, and, finally, as the sun set, stood on top of the highest mountain in Tierra del Fuego, Mount Darwin.

"Well, we knocked the bastard off!" said Jack.

We took turns photographing each other, ice ax in the air, with the National Geographic Society flag unfurled. We worked our way back down to the surface of the ice cap. As soon as we hit the flat, fatigue set in. Night fell; we turned on our headlamps and retraced our steps methodically. Hour after hour, Jack pulled me faster than I wanted to go. There was nothing to do but slog it out, splitting the endless darkness with the small cones of light radiating from our foreheads. Bored by the monotony and fatigued by the drudgery, the active levels of my mind ceased to function.

As was my custom to block out pain and fatigue, I retreated to the recesses of my inner mind. As if to help me focus on a center of being free from the physical dreariness I was enduring, my mind lit up in colored hallucinations that fascinated me. Bright sparkles of light traced their paths before my field of vision and gathered together in a sphere like a swarm of fireflies. I focused on the swarming light and forgot the pain of lifting one foot after the other. I remembered an unusual neurological disorder. A patient continuously felt the pain of being frostbitten even in warm conditions. A specific chemical, activated in nerves in the skin layer, transmitted an electric signal to the brain activating the pain. Now, that's misery, never to be able to escape from pain. As if on automatic pilot, we trekked on willpower alone, finally reaching camp at one o'clock in the morning; we had been walking for eighteen hours straight.

I thought that was that, mission accomplished; "we'd knocked the bastard off," to use Jack's expression. But Jack wanted to stay longer. "Lots of virgins to bag around here," he said, meaning unclimbed peaks that would go into the record books.

Not interested in bagging virgins, I stayed in camp and melted snow for tea and instant mashed potatoes. For three mornings, the others left me alone with the barren ice, with not a jet contrail, a satellite, or a high-flying condor. Gentle breezes tugged at the red nylon tent. My mind went into a peaceful state; there was nothing to do but wait. On the third day, Jack, after returning from a climb, noticed dark clouds boiling in the snow peaks toward Antarctica. I crawled out of the tent to greet my comrades on their return. In the west, the pyramid summit of Mount Sarmiento was bloodred in the setting sun.

"We'd better get cracking to base camp." Jack pried off the rawhide bindings of his snowshoes with the tip of his ice ax.

"We're almost out of food anyway." I was aching to see the green of a tree or, for that matter, the black of a rock.

"We're out of Almond Rocas. It's definitely time to leave," Peter said, only half joking.

In satellite photographs, Tierra del Fuego appears as a tiny finger of islands jutting off the end of South America. Over a thousand miles to the west, in the tropical waters near New Guinea, a collection of warm-air zephyrs had coalesced and risen in the atmosphere. The Coriolis force spun the rising marine air, laden with water vapor from the warm ocean, into small cyclones and magnified the flow. The steamlike water vapor, condensing into misty drops, released enough heat to push the air mass higher. The warm, moist spinning air grew and expanded over the southern ocean. Huge streamers of clouds spiraled in toward the center of the growing wave cyclone, approaching speeds of a hundred kilometers an hour. Like an inverted comma, the cyclone filled the Southern Hemisphere with white clouds, blotting any view of the water. In the night it crashed over the Darwin Ice Cap like a tidal wave. I was woken by the sound of the tent zipper. Jack stuck his head out. I remembered I had dreamed the wind howled underwater; a storm of whirling water caught up dolphins and threw them onshore. The storm of my dreams merged into the roar of a fierce blizzard that battered the tent membrane and rocked it in gusts.

"Holy shit!" Jack yelled above the sharp snaps of the tent fabric, and turned back to us, one side of his head plastered with snow and a look of panic on his face. "I can't see a damned thing out there. Let's get out of here!"

"I told you," said Peter. "Now we're in for it."

The tent collapsed and knocked me over as I rolled up my air mattress. I pushed myself up and wrestled my air mattress and sleeping bag into my pack.

"Let's get our bearings," I yelled.

Jack brushed the snow off his head and pulled the air photo out of his pack. We huddled together and studied the photograph. We were in the center of the ice cap with glaciers radiating out north to Admiralty Sound, south to the Beagle Canal, and west to Seno Hyatt, the location of our base camp, where we had left the Zodiac.

"We trek a mile due north." Jack wiped the snow off his hair. "Then we turn west down the glacier to Seno Hyatt. Roger, you carry the compass, and I'll pace the distance."

"Let's not end up in Admiralty." Peter pointed to the glacier leading to the north. "We'll never get down that glacier, and what good would it do if we did?" We'd end up a hundred miles from our base camp.

I followed the others out of the tent, and as soon as my head emerged, the hood of my parka assailed me with machine-gun-like reverberations that changed rhythm with the great streamers of snow that swept over me. The storm itself screamed like a banshee, wildly changing pitch and direction from side to side. Our goggles fogged up from our breath, and we could barely see. The stinging snow caked on our faces, melted on our skin, ran off our mustaches, where it froze into rows of grotesque icicles. The storm, getting worse by the minute, scooped up the tent like a kite. We chased the flapping corners, threw ourselves onto the orange nylon fabric, and fought to roll it up.

As soon as I shouldered my pack, the wind rocked me. I leaned against Jack's pack to steady myself.

"Now, look what you've done!" screamed Jack, as angry as the wind.

"What?" I asked.

"You bent my pack frame." He pulled on the frame to straighten it out.

We roped up, Jack first, Paul second, me third, and Peter brought up the rear.

"Can't see fuck'all. Keep us north." Jack handed me the compass and brushed the snow off his face and goggles.

The others disappeared into the storm, and I was alone in whiteness so thick I couldn't see my own feet. Streamers of snow snaked between my legs with awesome ferocity. I stumbled forward and tripped over a snowdrift. The quivering rope disappeared into the storm without touching the ground. Only the impatient tugs that pulled me forward reminded me of the others on the rope.

The first time the wind knocked me down, I was surprised. I pushed on my ice ax to get up; the wind pushed me down like a massive forearm every time I tried to regain my feet. Then seemingly out of nowhere Peter shouted above the roar of the storm and the staccato flapping of my hood. "I told you this is the worst weather in the world." He reached down to help me up, looking like a Norse god with ice glazed over his eyebrows and beard.

The rope suddenly tightened from in front. "Goddamn it!" I winced, and pounded the snow. "Wait until I get up, goddamn it!" The rope yanked at me again.

"Why the hell doesn't Jack stop?" I yelled.

"He's out to prove himself." Peter's bright blue eyes were calm, even amused, in the center of the screaming cyclone.

"I'm scared!" As I struggled to my feet, I imagined the wind rolling me about like a cat swiping at a mouse.

"Of course, but it's fabulous!" Peter held me under my arms.

"It'll kill us." I yelled back.

"Perhaps!" Peter raised his arms and parried playfully at the storm with his ice ax.

I opened the compass and brushed off the snow. "Still works." A tug of the rope yanked me into the white wilderness.

I was the most terrified of the four of us. Perhaps for that reason, I was knocked off my feet over and over. I was getting tired, and Jack was tired of waiting for me to get back up. I felt helpless, like a piece of jetsam, but I didn't want Jack to know how scared I was. We trekked on, fighting the wind for balance, one foot at a time, meticulously and monotonously. The even pace and steady whiteness mesmerized me. I indulged in the trance and retreated to the warm place of fantasy sheltered from reality howling around me. The field of sparkling dots filled my vision and formed a repeating pattern like a neon kaleidoscope.

A gust of wind pushed me forward and my strides stretched out. We were descending.

The turnoff to Hyatt, I thought. I pulled on the rope with both hands, but Jack didn't stop. I pictured him stomping his snowshoes one after the other into the drifts with headstrong determination. Without stopping, without eating, we trudged in the trail Jack stamped out. I was confused, hoping that he was right, that I was wrong, that we weren't getting more lost with each step. Then with great clarity of mind, I was sure Jack was pulling us down the wrong glacier.

The storm turned somber, with no hint of brightness in any direction. Outside the storm, the sun must be setting, the condors were hunkering down on the cliff faces, and soon the dawning stars and galaxies would appear stretched out to infinity. Here, the wind sucked the warmth and hope out of me; I became numb to my own emotions. Layer after layer of feelings and thought peeled away until I was functioning automatically from the core of my limbic system.

I suddenly came up on Jack and Paul, standing beside a dark shadow that yawned below the waving streamers of snow. I looked down at the gaping mouth in the snow.

"Crevasse?" I yelled.

"Yeah, give me a belay; I'm going down." Jack turned, shouted back, and made a gesture like pulling rope with his mittens.

"There aren't any crevasses here!" I screamed, sweeping my arms over the snow surface.

Jack looked at me as if I were crazy. "Don't worry. I'll get us out of this!" He made a clenched fist, a promise of victory, and disappeared into the crevasse. Peter came up as I belayed Jack.

Camp in a crevasse! As headstrong as Jack seemed, I did have to admire his fortitude at climbing around inside the crevasse to look for a flat piece of ice to pitch our tent on. At least the crevasse would block the wind.

Jack returned. "Nothing! We'll have to camp here. Make a snow wall!" He made a circling motion with his arms. We dropped our packs and anchored them to ice axes.

Jack cut snow blocks with the snow shovel, and the rest of us stacked them igloo-style. When we pulled the tent out of its sack, it flapped furiously and whipped us with the guy lines. We collapsed on top of it and got the corners pegged down. We fit the frozen aluminum poles of the frame together, pulled the tent up taut, and flung our packs inside to hold it down. Before Jack crawled into the tent, he made a fist at the wind and yelled, "Screw you, wind!"

After we blew up our air mattresses and unrolled our sleeping bags, Paul pulled out the bag of provisions. "We're out of jerky, but we still have potatoes; how about Mexican?" Paul pulled the Primus out of his pack.

"Yeah, something hot." I was exhausted.

Paul set a pan of snow on the stove; the pleasant sputtering sound and flickering blue light distracted from the memory of the storm. The tent gave a tremendous heave, snapped back with a gunshot report, and blew out the stove. Darkness and the storm's roar filled the tent. I felt desperately cold.

"Goddamn it!" screamed Paul. He pounded the floor of the tent with his fist and groped for a flashlight in the dark.

"Imagine if we didn't have the snow wall." Paul struggled to relight the stove.

I crawled into my sleeping bag.

"What if we're on the wrong glacier?" Peter pulled his sleeping bag around him. The moisture from our breath froze on the cold nylon and coated the walls with frost.

"This is the right glacier. Drink some tea. You'll feel better tomorrow." Jack pulled the air photos out of his pack, rubbed the stubble on his chin as he studied them, and handed them to Peter. "How much fuel?"

"A week," Paul said.

"Food?" Jack asked.

"These are the last potatoes, no jerky, no oatmeal, about two days of bittersweet." Snow caked on the outside of the cooking pot melted, slid toward the flame, and flashed into wisps of steam.

"Here we'll starve or die of thirst. In four hours we could be at base camp," said Jack.

"Or four weeks to base camp—if we make it through the crevasse field." Peter held out the air photo and pointed to a crevasse field on the large glacier flowing north to Admiralty Sound.

"There are no crevasse fields on the glacier we came up." I pointed to our route on the air photo.

Jack looked at me, one eyebrow raised, and patted his sparse blond hair. "I'll be damned. How long do these storms last?"

"Sometimes two weeks, maybe three," replied Peter.

"Oh, great!" Jack replied sarcastically.

"Who knows," I said. "If we stay here, it might clear up just enough to get our bearings."

Paul handed out cups of tea and turned off the stove, darkness filled the tent, the nylon walls shivered, and the wind outside moaned. Without the snow wall, the tent would already have been torn to shreds.

I slept fitfully and woke up when Jack pushed through the tent flaps. "Worse than yesterday; can't see a damn thing."

It was Peter's turn to go out for a leak. "I know climbers who have tubes leading out from their sleeping bags."

Paul lit the stove and melted a pan of snow. "Watch where you piss, Peter. I'm getting the drinking snow from that side of the tent."

After I found my socks, I took my turn outside. The urine splashed about in a yellow patch next to those of Peter and Jack. The yellow of the snow reminded me of a pale sun. But there was not even a memory of sun in the thundering white sheets. I crawled back inside my sleeping bag; Paul handed me a cup of tea and a semisweet chocolate bar.

I cradled the cup in my hands and let the warmth flow up my arms. When the tea hit my stomach, I felt terrible heartburn and a pain in my stomach. Damn wind's giving me ulcers.

A savage gust dive-bombed the snow fort from straight up and flattened the tent over us; then it snapped back into shape, like an umbrella turning inside out.

"That was close," said Jack. "Fucking wind."

"The wind is not against us." Peter chewed his chocolate slowly and meticulously.

"It sure is trying to knock the crap out of us!" Jack looked at Peter, one eyebrow cocked.

"We're incidental," said Peter.

"Of course we are incidental. So what? Tell me, what do we do?"

"Talk to it," suggested Peter, only half joking.

"Bullshit," said Jack. "What do you mean 'talk to it'?"

"The Yahgans believed it was possible." Peter shrugged his shoulders.

"Right, and they're all dead," said Jack. "A lot of good it did them."

"Not because of the wind," answered Peter.

"Well, I believe in action," said Jack. "We're wasting time sitting here. I say we ship out and move down this glacier."

"Jumping out of the frying pan into the fire," I replied.

"Fire, fire, everywhere, and we can't see beyond our feet. Land of fire, bullshit! Land of frostbite and hypothermia," ranted Jack.

"I say we sit here until it clears," said Paul.

"Could be weeks," retorted Jack.

"Hey, if we go down the wrong glacier, it could be months." Paul held out the teapot. "More tea?"

"No. Hey, I'll talk to the wind." Jack stuck his head out the door. "Screw you, wind!" The wind snipped at the flap of the door and the zipper slapped Jack in the face. He slunk down in his sleeping bag.

"Not like that," said Peter.

I cringed when Jack cursed the wind. Rationally, I knew that it didn't make any difference. Yet the more I felt beleaguered by the wind, the more I felt as if it were alive. But I agreed with Peter: if it was alive, it didn't notice us.

"Let me tell you the story about Martin Gusinde, the Argentinean anthropologist," said Peter.

"Why not? We got the time," said Jack.

"Martin Guisinde lived with the Yahgans from 1922 to 1924. He was one of the few Europeans who studied the Yahgans without prejudice, and he reported that the Yahgans believed that death is the separation of the spirit from the body, but they knew nothing about the fate of the spirit after death except that it goes on living forever." Peter stopped and took a sip of tea.

"The Yahgans believed that dream figures are actually souls of spirits, ancestors, and even animals that present themselves as living manifestations. They believed all things had spirits and all spirits are interconnected.

"Almost every morning, one of the adults in a family hut would relate his dreams; sometimes they would wake each other up in the middle of the night to relate a dream, or even call out to all the neighboring huts. Each individual interpreted his own dream, and the other villagers accepted it without question. They each remembered proofs of their dreams' reliability and pointed to them over and over again. Almost any Yahgan could point to the reliability of a dream.

"In the summer of 1923, a Yahgan woman, Nelly Lawrence, told Guisinde of a dream she had the night before: 'A few families came from Mejillones [on Navarino Island] and told me there was universal mourning over there. Today someone will surely come from there and tell us who has died.' Trying to cheer her up, Guisinde told her that dreams are meaningless.

"At noon a messenger arrived from Mejillones and announced, 'Last night two women died over there among us.' Nelly came to Guisinde, 'So you see my dream of last night, which I told you about this morning, came true,' she said with sad earnestness. 'Whatever I dream always comes true, no matter how distressing. Now we must fear that still more people will be taken, and soon the last of my countrymen will pass away. Then I shall be the only one left and must feel the hard fate of my countrymen. I, myself, will be here but a short time, and with me the last survivor of the Yahgan will sink into the grave.' She was right." Peter sipped more tea.

For five days the storm didn't let up for a minute. We ran out of food, and in six more days, we would be out of fuel for melting snow. Each day we spent more time sleeping, and only with great effort did we go outside to piss. The urine splashes in the snow got as red as the setting sun. The wind was sweeping the fluids from me. I got terribly constipated as dehydration set in and, with it, a great feeling of lassitude. What was the use? We were hopelessly lost in the most forlorn corner of the earth, racked by a hellish wind, without hope of rescue.

At night I dreamed the boiling gray storm turned into a circling cyclone of condors. I was entombed in a large block of glowing blue ice about ready to fall into the fjord. One condor winged over, scrutinized me carefully, and announced, "He's not dead yet."

In the morning, Jack pushed himself up slowly on his elbows and blinked his eyes at the quiver of the tent. The tent shuddered; the stove sputtered and went out. Paul held his head in his hands and collapsed back on his sleeping bag, heaving for breath. After resting for an eternity, he pushed himself up on one elbow and relit the stove. It blew out again, and again, before he got it lit. Dark shadows rimmed Peter's eyes, yet they burned like bright coals.

Paul handed out cups of warm water.

"Time to go." The fabric of his sleeping bag muffled Jack's voice. "We're losing energy. Time to head down."

"No, look at the photos." I sipped at the cup of warm water. "The rock cliffs. We can test our location by trekking west. If we run into those cliffs, then we are on the wrong glacier."

"Good idea." Paul spoke as if his mouth were full of cotton.

"A rational approach." Jack spoke slowly in a hoarse whisper. "I like that."

With only the greatest effort, in slow motion, as if moving underwater, I climbed out of my sleeping bag. Peter laboriously stuffed his sleeping bag into its nylon sack. Jack found his wool socks in the bottom of the sleeping bag and pulled them on. I pulled on my thermal boots. Since the laces were frozen, I had to thaw them out by rubbing them with my hands. We rolled up the air mattresses, stopping for frequent breaths. I bumped into the tent wall; frost rained down on me, and trickled down my collar. One after the other, we crawled out of the tent, heads bowed to lessen the impact of the stinging snow. Like swimmers fighting ocean currents, we fought the wind for possession of the tent and, inch by inch, got it into its sack and lashed on Jack's pack.

The storm pounced on us as soon as we left the snow fortress. We struggled with our snowshoe bindings and staggered into the white fury on a heading that we thought would bring us nose to nose with the rock cliffs that would test our position. I was much weaker than when we had stopped five days before, my mouth was parched, and my stomach in terrible pain. The snow fortress dissolved into the whiteness behind us. Jack broke trail like a drunken sailor stalking the deck of a wave-tossed ship.

"To the left," Paul yelled hoarsely to Jack.

Cradling my ice ax in my arms, I pushed against the wind. The monotony rolled over us. My leaden legs mechanically trudged in the shadowed snow prints. The even pace, the constant pain, and the steady rage of whiteness mesmerized me. Retreating into the recesses of my mind with each step, I sank into a trance, a refuge beyond conscious recognition of the horror of the blizzard.

Revelation crept from the warm corners of my mind. The dots of light with comet tails took on tinges of color like lead crystals and formed a lacelike veil of light. Chombi said, we relive our worst fears until we no longer fear them. Can't stop fearing the wind. Fuck, who cares? Sleep, just let me sleep. In Antarctica we were told to not give in to sleep. Ice cream, yes, ice cream warms you up. My wrists, at the gap between my parka and my mittens, were numb with cold.

We started up a slope of soft snow, and I sank in up to my calves. I stumbled and stepped on the rope.

"Goddamn it," Jack moaned, and pulled the rope out from under my snowshoe.

We climbed uphill for an hour, until it was no use. We hadn't found the cliffs and were more lost than ever. I stumbled and stepped on the rope again. This time it pulled Jack backward into a pile of loose snow. He pounded his fist into the snow. "Shit!" He struggled to push himself up on his ice ax.

"It's no use," I shouted, and dropped my head.

"Shit." He looked up with pleading in his eyes. "My feet." He pointed to his snow-caked boot.

Peter stumbled up and dropped his pack.

"Jack's feet. Frostbite," I shouted above the roar of the wind, and pointed to Jack's boots.

Paul pulled the snow shovel out of his pack, sectioned out a snow block, and lifted it. It exploded in a puff of powder that was swept away by the storm. We couldn't make a snow wall.

"Too soft," muttered Jack in despair. "Put up the tent anyway."

We stomped down a level patch, unfurled the tent, fighting the wind for the corners, and anchored them with ice axes. We fit the frozen poles together and as we stretched the agitated nylon in its hoops, the framework flexed and snapped back.

"Inside! Before it breaks!" Peter yelled. We crawled in and braced our backs against the corners. The throbbing tent rocked us back and forth.

"How long before it rips?" I asked.

"Who knows," answered Paul. "Once it does, we're finished."

Jack anxiously brushed the snow off his boots and tore away the socks. His feet were blue and stiff.

"Put them in here." Paul unzipped his down jacket and cradled Jack's feet inside. The tent knocked them sidewise. They pushed themselves up again. It was useless to light the stove. I felt numb and cold all over, and struggled to hold my corner against the storm.

I was beyond despair, grief, beyond any emotion or any instinct. I lost consciousness of the storm, the flapping tent, my hunger pangs, awareness of my body. I retreated into internal thoughts. Frost falling off the nylon as if it's snowing inside. Only a miracle can save us. Strange, a mosaic of diamonds, together in a shimmering globe of light. The orange skin of the tent gone. My mind collapsed. *I am about to die, in this place on this day.* Need to contact Jill. Jill. Everything is finished.

Yet there was another part of me watching as if through a telescope from another universe. It was like those movie scenes of mountain climbing where the audience looks down from a higher vantage point on the climber ascending a steep ridge. Our mind is willing to engage and accept the idea that the climber is overcoming incredible obstacles and difficulties, without focusing on the obvious fact that a cameraman had already climbed those obstacles with a heavy camera. The reality was that at least two individuals were on the ridge; otherwise the audience wouldn't be able to look down on the climber. As the lower levels of my mind abandoned their sense of reality, I focused on the cameraman. The one who watched the action became the real me.

It was the lower levels of my mind that couldn't face the anguish of giving up life, and once they were abandoned, there were no misgivings, there was no despair; on the contrary, there was peace. I became sure, focused, calm, and remote from my surroundings. Then I heard Peter's voice, as if far away and below.

"There's one thing we might try," he said slowly, deliberately.

"What? I'll try anything." Jack didn't look up.

"Yahgans believe everything has spirits, and the spirits are connected and can be contacted."

Paul took a deep breath from his inhaler, thumped his chest, and gasped for air.

"Imagine a break in the storm. Make a picture in your mind of Mount Sarmiento." Peter closed his eyes with calm deference.

No sooner had Peter finished than I was back in my own inner space whirling with light. My mind spontaneously generated an image of Mount Sarmiento in vivid color. Instead of experiencing a dream that rolls through the mind beyond control, I consciously and effortlessly participated in the formation of the image, one stronger and steadier than a memory. Peter, Paul, even Jack, held their heads in their hands as if in prayer.

"But without doubt, no second thoughts, send the image to your subconscious where it connects," Peter chanted.

For my part, I was too tired, too convinced that we didn't stand a chance to offer any resistance to Peter's suggestion. Visualizing a mountain through a gap in the mists was a wonderful escape. Mount Sarmiento, framed by layers of soft clouds, became a clear, still point in the center of my imagination. I could float into the scene and never come back. I lost track of time, or perhaps it slowed to a stop. I didn't care if I did come back. The clouds of Sarmiento engulfed me, and I couldn't tell if they were real or part of my imagination.

I was unaware that Jack had pulled on his boots and crawled out of the tent. When he yelled, "Well, I'll be damned!" I opened my eyes. I had no idea how much time elapsed. The tent was calm and full of light, no flapping of the fabric, no roar of the wind, no frost falling from the walls. "Come out here, you guys."

The tent didn't push in when I crawled out of my corner. I stuck my head out of the tent. The blizzard had stopped, and the clouds had lifted slightly off the surface of the snow. I stood up and looked over the ridges to the west.

"Look at that!" Jack pointed to the light patch of clouds with Mount Sarmiento fading in and out of the curtains of mist, exactly as in my vision. I have never seen a more brilliant pure white than Mount Sarmiento that day.

"Oh, wow! A miracle," said Paul with a faint smile.

"Fabulous," said Peter, and his eyes brightened in the shadows of his face.

"A coincidence," said Jack. "I mean, a good coincidence."

Under the bank of clouds, we could see the black cliffs.

"We're a half mile south from the cliffs," said Peter. "We must be on the wrong glacier."

"I'm going to check it out." Jack put on his snowshoes and plodded up the ridge.

"It's a waste of time. Mount Sarmiento is behind the ridge. It should be straight down the glacier." Paul crawled inside the tent to pack up.

"He doesn't want to admit he was wrong," said Peter. By the time Jack returned, we had packed up camp.

I felt exuberant, light-headed, and free of anxiety. Fate had pulled me through again; there was a God. Though still hungry and tired, I was reborn to a clear energy that had been present but veiled all along, and able to take up the trek again.

We retraced our steps, rounded a ridge, and descended into the valley of the Hyatt Glacier, protected from the brunt of the storm. A ceiling of swarming clouds closed over us, and we were once again socked in.

We broke up into two ropes. Paul wove back and forth across the glacier, and Jack barked commands at him. To the right, now to the left. We climbed the ridge above the icefall. Night fell, the wind picked up, and we were forced to make camp without reaching base camp.

The morning dawned dark and dreary. We were hungry and tired, without the energy to get out of our sleeping bags.

"Oh, man, I'm too tired to start the stove." Paul pulled out his inhaler, tugged on it, and threw it on the floor of the tent. "Shit, it's empty."

He pulled the first-aid kit out of his pack, searched through it, and flopped back on his sleeping bag. "No more refills." He rolled on his side, wheezing for breath.

"Let me look." I lethargically searched through the first-aid kit and pulled out a syringe.

"What's this for?" I asked.

"Demerol. In case someone breaks a leg," said Jack

"This?" I held up a bottle of yellow pills.

"Benza-something."

"Benzedrene?" I opened the bottle and held a pill in my hand.

"Energy for an emergency," said Jack.

"Energy for an emergency? This is an emergency." I swallowed the pill.

"Wait! They can drive you insane." Jack grabbed the bottle from me.

"Now you tell me." I collapsed back into my sleeping bag.

I dozed for half an hour, then woke up with a tremendous surge of energy. I grabbed my socks, pulled them on, jumped out of my sleeping bag, stuffed it in my backpack, tore open the tent flaps, jostled my pack out, and looked up at the sky. A current of clouds eddied overhead. The storm still raged around the peaks above us.

Jack looked at me, nonplussed. "Where are you going?"

"Base camp." I felt jolly, full of pep, and let out a yodel.

"It's three miles of snowfields and rock cliffs." He groped anxiously for his socks.

"No problem; I'll make like a condor and fly." I flapped my arms and hoisted my pack to my shoulders.

He pulled on his boots and threw a bunch of things in his pack. "That stuff's driving you loony; I'm coming with you."

"Suit yourself." I bounced the pack a couple times, tightened the waist belt, and scanned the slopes below. The veils of mist opened and closed with alluring glimpses of the fjord. I loped off, taking huge strides, floating a long way before my feet struck the snow, flying free. I twirled my ice ax as if it were Charlie Chaplin's walking cane. Half skiing, half jumping, twisting to break my slide, I bounded down the snow slopes. I felt totally at home in the world; everything was beautiful; no pain, no suffering. Death? What was I thinking? What a profound difference a yellow pill can make. I saw a dark shape in the snow and pulled up short.

"Look, Jack!" I turned back to him. He loped along, crunching the snow ten yards behind me

"What?" he asked, irritated.

"A rock, it's black, black rock. Have you ever seen such a beautiful black rock?" It glistened with mica and feldspar in the dull gray light.

Jack looked down at me, hands on his hips. "All right, you found a rock; let's go."

"No! It's warm and black . . . Jack." I removed my mittens and rubbed my hands over the rock as though it were a campfire. Then I spotted something even more appealing.

"Look down there." I raced down the snowfield to a puddle of meltwater.

"Water! I found water." I drank, splashed water on my face, brushed my teeth with my finger. Liquid water and the smell of fresh clay had become great luxuries.

Jack scrambled down the snowfield, knelt down, and drank from the puddle.

Then I saw something still more astounding, and dashed off, flying down the scree slope. My boots flew from boulder to boulder as if in slow motion. Without thinking, my brain accurately gauged the distance and impact of each stride while I was still in midair; there was not the slightest chance I would trip and break my leg. I knew I would make no mistakes. The patches of snow between boulders gave way to patches of grass that grew in size and merged into larger ones. I felt unusual softness under my feet.

"Grass! I found grass!" I yelled back at him. He was nowhere in sight.

I rolled about, stretched, and swam across the grass. The herbal smell intoxicated me all the more, much better than the smell of white gas fumes and old sleeping bags. Jack caught up and collapsed in a heap on the grass.

"Isn't this heaven?" I asked.

"Don't go running off. Wait! Goddamn it."

I started running, landing with a great drive of my boot heel, bouncing on the turf, five feet, running, then taking another yodeling sail down the sides of the world. I saw the silhouette of a beech tree in the mist.

"A tree, Jack, I found a tree." I darted toward it.

"Hold on." Jack ran after me.

"I am invincible!" I grabbed a branch and swung back and forth. Jack ran in front and spread out his arms. "Watch out, goddamn it; there's a cliff here."

"It's impossible to fall off mountains." I swung down to the grass.

Dark shadows in the wind rose up out of the mists, perhaps condors riding the updraft along the cliff. I started running, arms outstretched.

"These condors like me," I said, convinced of my opinion.

"Wrong, they're waiting for you to fall off the cliff and kill yourself."

After an easy half-hour jog, we reached the ledges just above base camp. Jack, deciding I was no longer in danger, or a danger to myself, gave up and fell behind. The smell of the salt air and tangy kelp penetrated memories of the sea in my brain core, already reassembling the patterns of everyday cravings and instincts that I had lost in the storm on the ice cap.

I trotted across the cobble beach to the cache, tore open duffel bag after duffel bag, and threw everything in a heap on the ground.

"Why is the thing I am looking for always the last thing I find?" I was rummaging through the last duffel bag when Jack came up.

"What the hell are you doing?" asked Jack.

"Looking for these." I triumphantly held up the pink carton of Almond Rocas. "Oh, man, do I have a terrible case of the munchies."

"Go easy, you're not used to food." Jack tore into a pile of cheese wedges. I couldn't sit still. I filled up my pack with food, grabbed an inhaler for Paul, and headed back up the cliffs. I met Paul and Peter halfway down and handed the inhaler to Paul, his face ashen his chest caved in.

He sat down on a rock and tugged on it. "Aaah! I can breath again. Thanks." A pink color dawned on his cheeks.

Peter and Paul ravenously tore at cheese wedges, raisins, and cashews.

"And for dessert." I held out a handful of Almond Rocas. "There are condors waiting for you, my friends," I said.

"Oh, boy," said Paul, and he took another toke on his inhaler.

We pitched the tent on the cobble beach next to the Zodiac. I prowled among the rocks on the shoreline, found some mussels, and pried them free with my ice ax. Paul set up the Primus on the beach, boiled water with kelp strands, threw in the mussels, and as soon as they opened their shell, he added some Italian seasoning. They were delicious. Now that we were truly safe from the hellish wind, my mind split into a spectrum of mundane cares, worries, and preoccupations. Would we make it out through the pack ice? How would we get back to Punta Arenas?

The next day we poled our way out through the ice floes, and when we came into sun near the entrance to Seno Hyatt, we made camp. Cormorants preened happily on white-frosted cliffs and spread their wings toward the sun like solar collectors. Jack declared a day off; I decided to climb a rocky summit behind camp.

As I climbed higher, the rain forest opened up into moss gardens and groves of bonsai-shaped evergreens manicured by the Fuegian storms. Here was a true wilderness, untouched by man, yet beautifully laid out with lawns of trimmed grass and the smooth trails of deer.

I crested the rocky summit and gazed down at the fjords. Along the shorelines, the waves shoaled into silver ribbons outlining the fingers of sparkling water. The day was calm and silent. To the west, Mount Sarmiento's summit stood dazzlingly white against the profound depths of space. I had never seen such a paradise. This was not the harsh Himalayas, or the somber Amazon.

The expansive view filled me with a power and peace that transcended all previous memories of ecstasy; I was as complete in that moment as I ever could be in countless stellar lifetimes. I understood why the Yahgan shamans believed the mountains, trees, and animals have spirits that cry and call to one another. I wanted to stay, build a beechwood hut, and live off the land with a soul mate in this Garden of Eden. With that thought, the feeling of ecstasy slipped away.

Sitting down on a boulder with my climber's ice ax across my knees, I tried to regain the moment. But the harder I tried the more elusive it became. I focused all my attention on the rugged landscape. Not one road, airport, or power plant marred the expanse of land that stretched north past the Strait of Magellan. The remoteness of the location, the lack of human devastation, had undoubtedly facilitated the extraordinary feeling.

I heard a mysterious sound in the crags beneath my feet, like wind through the treetops. But there was no wind—and no trees.

Five black shadows swept out from the rock pinnacles below me. I jumped up, startled, and grabbed my ice ax. Then I noticed white collars on featherless necks. Condors! I thought. They're condors! They floated without moving their wings, long feathers on the tips rippling like the wind in the still air.

Naked red heads, small and out of place on the magnificent expanse of black feathers, craned over the terrain below, seeking out carrion. The featherless heads enabled them to scavenge decaying cadavers without infection by bacteria, the lethal cadaverites, that decompose flesh.

Missionary reports state that the Yahgan Indians dismembered the bodies of the dead and fed the corpses to condors. While repulsive to our culture, the practice reflected their absolute conviction that the soul endures independent of the body. We prefer to bury the dead, but if we actually watched the work of cadaverites and a host of other parasites, we might find it no less gruesome than the work of condors.

In the distance, the condors joined up with several dozen others, soaring up like a thunderhead above a hot spot on the landscape.

Yes, sheltered by harsh winds, steady rains and mountain

walls, Tierra del Fuego will persist, its forests uncut and its golden grasslands unplowed. The thought made me jubilant, and I started down the mountain.

I soon stopped for a lunch of chocolate and cheese. The bright sun and warm air instilled me with a luxurious drowsiness. I will miss this paradise, I thought. I sprawled out on a patch of moss and dozed off. A thick shadow passed over the sun. I opened my eyes to see a condor, dark wings twelve feet long, hovering only an arm's length away. The long distal feathers arched up and sifted the air, the tail twitched in rhythm, holding the condor in place like a kite tethered to a string. As it swayed from side to side, the neck stretched in the opposite direction, and the bald head scrutinized me first with one eye then the other.

I should have been afraid, but instead an unusual peace overwhelmed me. Looking into the condor's luminous eyes was like looking through darkness to a light from a parallel universe. An indescribable energy passed between us; I felt a strange affection, intense, almost erotic. My feelings of love embarrassed me. My mind filled with confusion. Without thinking, I reached for my ice ax. With a twitch of its feather tips and the sound of wind, the condor soared into the vanishing point of the landscape.

A terrible loneliness swept over me and with it the realization I was incapable of living in the rapture of the wilderness. I turned toward camp saddened but sure of one thing: if I had a choice between cadaverites and condors, I'd choose condors.

The next day we started out on the long journey to Punta Arenas with no clear idea of how we would survive once we reached the open sea or how we would cross the Strait of Magellan. Fortunately for us, in the early afternoon we encountered a fishing boat anchored in the sound near Mount Buckland. The boat had just dropped off an expedition of Italian mountaineers led by Carlo Mauri. Jack arranged transport on the fishing boat back to Punta Arenas. We transferred our gear and hauled the rubber boat onto the deck. The next day when we pulled into the dock at Punta Arenas, Jack stood triumphantly on the bow like one of Magellan's captains returning to Cadiz and directed the unloading of our supplies.

The Chilean minister of defense had explicitly prohibited the proposed second leg of the expedition, to Santa Inés Island. Jack,

Peter, and Paul decided to go anyway. I decided to leave for my new job as a geophysicist in the rain forest of Ecuador, which I had accepted before I left. The four of us gathered together for a last dinner of crab and sangria.

"You're sure you won't come to Santa Inés with us?" asked Jack, cheeks flushed with sangria.

"No, I'm already late starting my job in Ecuador." I dabbed a crab claw in melted butter.

"You're a great route finder; where'd you learn it?" he asked.

"Picking blueberries with my mother." Another feather in the cap of *Route Tracker*.

"We'll miss our blueberry-patch route finder. We should've carried you down the mountain on our shoulders." Jack raised his glass in acknowledgment.

"Oh, no, it was Peter's idea of doing the concentration that saved us, I'm sure," I replied. "Paul, what do you think of Peter's miracle?" I stood up from the table.

"A miracle is as a miracle does." Paul stood up with a mysterious grin.

"What does that mean?" I asked.

"I don't have the slightest idea." Paul shrugged and gave me a hug. It was the beginning of a lifelong friendship.

"Take care. See you next time around," I said.

"I'm sure it was a coincidence," said Jack, standing up. "It was your route-finding sense that saved us."

"Yeah, like my route to South Mountain across SAC's main runway." I reminded Jack of our first adventure in Greenland.

Jack laughed, and we shook hands.

I turned to Peter and shook his hand. "I'm beginning to believe in miracles."

"No, not miracles. Concentration! That's everything. *Abrazo!*" We hugged.

We walked out of the restaurant just as the airport shuttle pulled up. I picked up my pack and duffel bag. "Enjoy Santa Inés. I'll be seeing you."

As I flew north over Chile and Peru, I reflected on the events in Tierra del Fuego. First Everest, now this. Extraordinary things happened on the edge of death. Though in this case, unlike most near-death experiences, I did not experience the event alone. It was more difficult to dismiss the experience as a hallucination of my own. With the possible exception of Jack, the other members of the expedition agreed something miraculous had happened. Apparently, in a moment of desperation, we had found a level of consciousness that allowed us to move the clouds and cause the storm to abate. It was the perfect demonstration of the Yahgan belief that all things are connected.

Yet I didn't really believe that all things are connected. Why do things seem separate some of the time and not at others? What is it that connects everything? Psychics speak of strange magnetic forces between people and things. But no such force had ever been detected or measured by science. Even though I had direct experience to the contrary, I could not replace the framework of normal experiences that defined my life.

Sometimes individual molecules of water or individual grains of sand behave as if they are connected, in wave shapes, for example. Ever since I was a child, I had marveled that wave forms appeared at so many levels in nature: waves on the ocean, waves of amber grain, waves of light, as well as waves of electrons. Perhaps all things in nature are connected. But how did that explain our ability to mentally move clouds on the Cordillera Darwin? The only rational explanation seemed to be that at some level we, as individuals, are connected remotely to other parts of the universe.

But how? If the universe is truly interconnected, why do we experience it as separate? Why do we feel lonely, cut off from the rest of the universe? The standard explanation I had heard was that loneliness is an illusion. That explanation was unsatisfactory to me, a cop-out.

Something extraordinary was going on, even though I didn't know what it was.

I stayed three years in Latin America, first as a geophysicist in Ecuador and later with Bethlehem Steel in Brazil. Between jobs, I engaged in mountain climbing and photography in the Andes of Ecuador, Peru, and Chile. Jill and I met for a week in Jamaica but soon drifted apart. In the meantime, I met Maggi, the love of my life.

CHAPTER FIVE

LETTING GO

It was May 1968. The Varig 707 jet from Rio, Bahia, and Belém pulled into the arrival bay at JFK Airport. As the plane lurched to a stop, the snap of seat-belt buckles traveled up and down the aisle; the passengers stood, stretched, and reached for the overhead racks. The remains of humid air, sweet as mangoes, spilled into the cabin as the passengers pulled down their hand luggage. I shouldered my camera bag, and when the stewardess cranked open the door, a tide of acrid air washed through the cabin. For three years after our adventure in Tierra del Fuego, I had lived in South America. This was my first visit home. As I walked toward the arrival gate, the sun-drenched faces from Rio and Bahia turned pallid, greenish, and bored under the glare of fluorescent lights and the aluminum panels of the escalators.

A glassed-in balcony, like one in an operating theater at a medical school, surrounded the immigration and customs area. What kind of vivisection was this audience anticipating? I searched for Maggi among the faces peering down. In her last letter she had said she would meet me and go with me. I wasn't sure what that meant. Things were far from right between us, and I doubted she would show up.

The immigration officer, with strong Mediterranean features under his white hat and black visor, scanned the passports with a bored officiousness.

"Next please." His voice was as crisp as the starched white collar that pushed up rolls of skin around his neck.

He took my passport without looking at me. "How long were you in Brazil?" He leafed through my passport.

"Six months," I replied matter-of-factly. I thought I saw Maggi out of the corner of my eye. When I looked up, it turned out to be a young man with long black hair.

"And what were you doing there?" He pulled a bound book of computer printouts from under the counter.

"I was working as an exploration geophysicist for Bethlehem Steel." I looked at him and set my camera bag on the counter.

"I see. Before that?" He leafed through the computer printouts, paused, and bent over a page.

"In Ecuador."

"I see." He ran his finger down the page. "And it says here you were in Chile before that."

"Right, Chile." If he already knows the answer, why is he asking me, I thought.

He compared a line in the printout with the numbers on the front of my passport. He laid my passport in a heavy metal paper punch and hammered it down with his fist.

"There!" He inspected his handiwork with pride and looked at me for the first time. "Mr. Hart, I have to inform you that your passport is no longer valid." Without changing expression, he handed it back to me, the word *VOID* stamped in perforations across the cover.

"What do you mean?" I asked, confused. "There must be some mistake."

"Contact this office in the State Department in Washington." He handed me a card. "Proceed to customs."

I couldn't believe it. "You mean I can't leave the country?"

"Not until you get a new passport at that office." He pointed to the card.

Trying to explain, I said, "We addressed the Chilean minister of defense as Señor Monkey Face. But that was a typo!"

Again I thought I caught sight of Maggi, but when I looked up it wasn't her.

"Proceed to customs." He closed the book and pointed across the lobby.

I shuffled off toward the baggage area, fingering the passport.

I had signed a waiver with the State Department after our expedition to Mount Everest; technically we had entered Tibet illegally. Tibet, after all, was controlled by communist China and therefore off limits to Americans. But I thought that had been cleared up.

There *was* an additional problem that arose during our expedition to Chile. Jack, Paul, and Peter had gone to Santa Inés Island against the minister of defense's wishes, and he had kicked them out of Chile. But I wasn't at Santa Inés with them. Nonetheless, the Chilean government included me in their sanction. Later, when I tried to return to Chile, I had been turned back as a spy at the Antofagasta border, but that was a mistake. Perhaps the State Department had somehow got wind of the Chilean government's action and had mistakenly blacklisted my passport.

I watched the luggage from the Varig flight pour under the rubber fence. My red backpack was nowhere in sight. Of course, mine will be last, I thought to myself.

I searched the faces in the arrival lounge for Maggi's. After being turned back at the Chilean border, I had hitched north toward Ecuador. I'd stumbled onto Machu Picchu, and there I bought a bamboo flute, now in the backpack, that seemed to carry the magic from that sacred place. When I was at my loneliest, hitchhiking the Pan American Highway on fishmeal trucks, I had met the Bronx-born Maggi, a Peace Corps volunteer, who had taken me to her home in the slums of Lima.

I stayed with her for two weeks in her adobe hut next to the sea. We became lovers. Our time together in Peru was out of this world. Brown pelicans cruised the updrafts along the sea cliffs like bomber squadrons. As we read by candlelight, the soft breezes filtered through the reed mats of her roof. She respected, trusted, and loved the poorest people in the worst slum in the world. She broke down in tears as she said good-bye to them, and I knew I wanted to be loved the way she loved them. Because she needed to be near her parents and I needed to stay in Peru for work, we broke up. But after she returned home to the Bronx, I realized I was in love with her. I wrote to her and asked her to meet me.

I watched the luggage ride around on the steel carousel. The mood of the airport was cold, aloof, and officious. The airline probably lost my luggage. I could be filling out forms in this aluminum-and-glass nightmare until morning. Finally, my pack slid out under the rubber fingers. I dragged it off the carousel, straightened it out on the floor, and shouldered it.

There was still no sign of Maggi. By the time I got to the customs line, I was feeling dejected and expecting the worst. With *my* luck, they'll strip-search me, I told myself.

But usually when I expected the worst, it didn't happen. Expecting the worst was a preventative visualization. Only the things I didn't expect, like having my passport voided, happened. In a sense it was my fault that my passport had been voided, because I hadn't expected it.

I concentrated hard on expecting to be strip-searched so that it wouldn't happen. I loaded my pack on the customs bench and pushed it toward the customs officer.

"Where're you coming from?" he asked.

"Brazil." I handed him my passport.

"Just voided, I see." He flipped through the passport.

"Yes, but it's a mistake. I'm going to get it straightened out in Washington."

I thought I saw Maggi sitting in the arrival lounge; it was a young woman who looked like Maggi.

"No drugs in Brazil, are there?" asked the customs inspector.

He looked me straight in the eye.

"Well, the shamans and voodoo priestesses use them. I don't. I have some prescription Valium. Ulcer." I pulled a small prescription bottle out of my shirt pocket and showed it to him. Ever since Maggi and I had broken up, my nerves were wrecked.

"Aha, here's what I need." He pulled out a slip of paper evidently placed in my passport by the immigration officer. He chalked an *X* on my pack and my camera bag.

"Go on through." He signaled a skycap.

"Okay, thanks, I don't need a porter." Triumphant in the success of my preventative visualization, I shouldered my pack and camera bag, searched the crowd in the arrival lounge, and headed to the exit. She won't be here, I told myself.

I slid through the automatic doors to the sidewalk and stared at the line of yellow cabs moving forward like a centipede.

"Roger! Roger!" I heard her yell over the din of traffic. I turned and saw her rushing through the crowd. It was really her. She carried an overnight bag. That was a good sign. She really was going with me.

"Sorry. I'm late." She ran up to me, breathless, in a white peasant blouse embroidered with red llamas over her breasts.

"I knew you'd show up." I dropped my pack on the sidewalk.

She dropped her bag, hugged me, but didn't give me a kiss.

"It's good to see you. How are you?" She pulled away and looked at me with grave concern etched on her ivory forehead.

"Great!" I looked back toward the immigration desk. "Well, pretty good. I mean this thing just happened. How are you?" I shrugged and looked back at her.

"Fine," she replied. "I need to talk to you."

"Okay, let's talk." A Gray Line bus pulled up to the curb. "What are we doing? Taking the bus?"

"Yeah, this one will take us to Port Authority." She picked up her bag.

We handed our luggage to the driver, who tagged it and slid it into the compartment under the bus. She stuffed the claims ticket in her handbag, took me by the arm, and led me to the steps up onto the bus.

"Are you sure you're okay?" She looked back over her shoulder as she climbed the steps in her plaid miniskirt with black hose.

"I had a thing with immigration." The more I thought about it, the more depressed I got.

"What kind of thing?" She slid into one of the maroon velveteen seats.

"Oh, they voided my passport. You know. I signed a waiver to get it back after Everest. Then the screwup in Chile. It's a mistake. I'll get it straightened out." I collapsed into a seat next to her and took her hand. She gently, but cautiously, squeezed back.

The driver climbed briskly aboard and pulled the door shut, the air brakes hissed, and we wheeled onto the highways of Long Island. I stared out the smoked-glass window. Compared to the narrow dirt streets of Latin America, the flying concrete ramps seemed massive, imposing, and menacing. Darkened cars whizzed in and out of the sodium-vapor light, driven by faceless creatures of the night. What acorn or chestnut could sprout in the chlorinated, hydrocarbon-oil-slicked soil under these flying overpasses?

What was happening to me? I was in culture shock returning to my own homeland.

On the streets of Latin America dogs rushed out to greet you, children played on the curb, and handicapped beggars camped on the sidewalks. These highways had no sidewalks, and if a stray pet, out of utter confusion, wandered onto the road, it would be mashed on the sleek chrome of five different bumpers before it hit the road.

"Stupid people should keep their dogs off the highway," the drivers would say. Without emergency lanes, it was impossible to stop or change a tire. When else in history have we been at the mercy of mindless drunks and countless cardiacs hurtling massive weapons of steel at us every day? The maze of concrete was imposed on me against my will, and it depressed me to see so many unaware of the threat.

Welcome home, I told myself; only a miracle can save you. My mood got worse before it got better.

"How can anybody live in this soulless crap?" I waved out the window. "Not a home for man, beast, or vegetable matter in sight." I let go of her hand and took the bottle of Valium out of my pocket, unscrewed the cap, and swallowed one of the pale yellow pills with the number 50 etched on it.

"What are those?" she asked.

"Valium. Ulcer." I put the Valium back in my pocket and held my hand over my stomach. "I feel ridiculous having an ulcer. It's not like I am a CEO making million-dollar decisions every day."

"You're dealing with a lot. Your family. Your country," she replied earnestly.

I took her hand and squeezed it, and she squeezed back sympathetically. I felt all right with her at my side, but I couldn't imagine what emotions and visions would surge through my brain once we were apart.

We ascended an overpass, and I looked down on lines of dark square houses with cones of pink-amber light in front of each of them. Like firebreaks, the highways surrounded the flat lawns and protected them from thicket, bog, field, and forest, fortresses against chaos. A little chaos never hurt anybody. Open our hearts, let it into our lives, I thought. Complexity and self-organization ride the coattails of chaos; diversity, that's what's important.

"Now I remember why I don't want to have kids." I looked at Maggi.

"What do you mean?" Her mouth dropped, and she let go of my hand.

"You work eight hours a day, hate your life, and fight like

sharks over every scrap of validation. Why subject a child to that kind of life?" I swept my hand dismissively over the urban scene.

"Well, you liked the kids I lived with in Peru!" She scowled at me with an edge of disdain.

"I did like them." I panicked when I realized how harsh her look really was. "I liked the light in their eyes." I sensed that her mood was lightening up. "The parents didn't resent them. That was different."

In the *barriada* where Maggi lived in Lima, the parents, even though they were poor, always had time for their kids, and the families watched out for each other's children. During the day the children, in rags, ran from house to house laughing. In Latin America, love had no strings attached. It lived and grew by itself, an end in itself, not something to be earned.

"My parents immigrated from Ireland to live in this 'soulless crap,' as you call it." She crossed her arms over her breasts.

"Well, compared to Ireland, this might seem like paradise."

Her jaw tightened, she crossed her legs and stared out the window.

"Oops! Oh. Uhm. I mean, I'm sure the Bronx isn't like this."

"Roger, I have something to tell you," she said with a forced evenness to her voice.

"What's that?"

"I can spend the weekend with you," she said slowly, watching me carefully. "But I can't sleep with you."

I didn't expect this, and the worst immediately came to my mind.

"Is there somebody else?"

She nodded her head yes, turned, and stared out the window.

"Funny thing. His name's Roger, too, Roger Whitehouse." She turned to me with a look of great concern.

"Oh, my God. I didn't expect this," I said.

"I know. I'm so sorry. I thought . . . I thought you were never coming back from South America."

"I waited for you to come back to Peru," I replied.

"Yeah, I know," she said. "But I can't just up and leave my parents."

"Well, I'm here now. Maybe forever," I said glumly.

Like the sun burning away a fog bank, the reality of my situation dawned in my brain. But instead of going ballistic and hysterical, I calmed down. Nothing worse could happen to me, so I had nothing to lose. It was the fear of losing something that made me nervous, and now I had nothing left to lose, so I was calm. My head slumped down. So that was it, that's why she's going with me, I told myself. She was letting me down easy.

"Where do you want to go?" she asked.

"Oh, I don't know. Who cares?" I was disorientated.

"Well," she said softly, "there's a peace demonstration in Washington this weekend."

"Peace demonstration?"

"Antiwar," she replied.

"Oh, well, I do need to go to Washington. Gotta straighten out my passport," I replied. "Is this other Roger . . ."

"Roger Whitehouse?"

"Yes, is Roger Whitehouse going?"

"No, he's a designer and he's redoing our office this weekend."

The off-ramps disgorged the flow of traffic like rivers into the upper canyons of Manhattan. The bus lurched into the avenues of polished granite, plate glass, and monolithic brick walls that disappeared upward into the ceiling of smog. Spray-paint graffiti crawled up the buildings like crustaceans in an intertidal zone and lent humanity to the cold foundations. A broken-down car with the hood up spewed steam through the grill. Car and taxi drivers alike blew their horns, yelled, and swore at the owner, who had his head under the hood.

"Out of the road, asshole!"

The taxi drivers sped angrily around the car and then jammed on their brakes when they came to a red light less than half a block away. The bus stopped at a couple of hotels, slid south on Fifth Avenue, and, at Forty-second Street, the ground cover of graffiti sprawled over trash cans, subway entrances, phone booths, sidewalks, lampposts, and traffic lights. Trash cans looking like mushrooms bulged out on top, the contents overflowing.

We pulled into a bay at the Port Authority bus station, grabbed our bags, and then stalked the expanse of the terminal, searching for the Greyhound departures to Washington. I grabbed Maggi's limp hand and clutched it like a life preserver.

Young people in bright clothes sat in clusters on the floor, strumming guitars, talking livelily, and snapping their fingers. Like Rio during Carnival, I thought. Some wore relics of army surplus clothing—boots, camouflage flak jackets, olive drab pants—set off with brilliant colors, an American flag here, a piece of tie-dyed cloth there, or Day-Glo-spray graffiti like that on the building foundations. Others dressed as iconic or TV heroes: Davy Crockett in coonskin hat, Kit Carson and Buffalo Bill in buckskin tassels, the Cisco Kid in ornate black sombrero, Zorro with whip and black cape, Chief Crazy Horse with headband and necklaces, Uncle Sam and Abraham Lincoln in tall hats. I stared at them as we walked past and caught snatches of songs. *"In the pine, where the sun never shines . . ."* Some smiled back and gave the peace sign. When someone said, "Peace, brother," I realized I had been staring and looked quickly away. Bob Dylan was singing "A Hard Rain's A-Gonna Fall" on a portable radio.

"There's a bus leaving for Washington in an hour," Maggi said, craning to see the departure listings. "Let's get our tickets."

"I caught me the end of an old freight train, and I never did come back."

"Good idea. Let's get something to eat."

We wandered east on Forty-second Street to Times Square, past porno houses and fast-food restaurants. We stopped and watched the bright letters on the old Times building. The stationary bulbs lighting up in succession created the illusion of moving letters. The face of a man on a large Camel cigarette billboard blew smoke rings, each one identical to the previous one. *"Give peace a chance,"* was written in small letters to the side of them. White marquees with flashing lights stretched up Broadway.

"Oh, look, an Automat," I said. "I always wanted to eat in an Automat."

"All right." Maggi dropped my hand and pushed the door open. Aluminum panels with coin slots and glass windows lined the walls.

"It's like walking into a vending machine." I slid two quarters into the coffee machine.

Older men in ragged war-surplus clothing and knit caps hovered over cups of drip grind on hard linoleum-topped tables. I took a sip of my coffee as we slid our trays past the food windows.

"How do they make coffee this bad?" I asked.

"Preservatives and incessant processing." Maggi shrugged. "It's the American way."

"But it's bitter, sour, acrid. It jangles my nerves," I said. "Brazilian coffee is smooth; it doesn't even bother my ulcer." I picked up some custard. "This will soothe the ulcer."

Maggi was the only woman in the cafeteria. When we sat down at a table, none of the men looked up. Probably junkies, I thought to myself.

"How's your custard?" Maggi asked. Then she jumped up suddenly. "Oh, my God!"

I turned to see one of the men at the table next to ours falling to the floor. None of the other men looked up. Probably an overdose, I thought to myself.

Maggi was to the man in a flash, lifted him under the arms, and helped him establish his balance on the chair.

"Are you okay?' she asked.

The man looked up, nodded his head, and waved her off. Every eye in the Automat followed her back to our table as she sat down.

"Maggi, I can't believe you did that," I said, staring at her.

"Why not? All I did was help him up." She looked at me and smiled.

"Nobody helps anybody in New York. You'll get mugged; everybody knows that." Actually I admired her courage: to show that sort of kindness in a cafeteria full of strange men.

"It's no big deal, really; I trust people," she replied.

The love I had felt for her in Lima flowed through me again. I reached for her hand. "Maggi, listen carefully. It's not easy for me to say this. I've never said it to anyone before, but I really do love you, like I've never loved anyone."

Her face contorted in anguish, and her eyes filled with conflicted emotion. We caught a Greyhound to Washington, D.C., and checked into a hotel, separate rooms.

The next day, row after row of somber figures swept down Ninth Street in Washington, fists raised in fierce demeanor beneath bright red banners. Chanting built to crescendo, echoed off the marble walls of the Justice Department, and reverberated in the maple trees. "Bob-by! Bob-by! Bob-by!"

"Who's 'Bobby'?" I yelled above the roar of the crowd.

"Bobby Seale, a Black Panther's been arrested," she yelled back.

As the marchers turned from Ninth Street south onto Constitution Avenue, the chants, trapped like bad feedback between the canyons of marble, echoed and reverberated on top of each other.

Maggi and I ran along the crowded sidewalk taking pictures.

Thousands of people chanting in unison generated trains of energy that gave me a rush of excitement and a surge of emotions I couldn't contain. Part of me wanted to join the protest, to be one of the group with purpose and focus.

"MACE! The pigs got mace!" someone shrieked, and started to run.

A wave of panic quivered through the ranks. The marchers pulled handkerchiefs up over their faces and looked anxiously over their shoulders, wide-eyed like frightened horses.

One of the marchers put his hand over my camera lens and said, "You take one more picture and I'll smash your camera."

I put my camera down and turned away. It hadn't occurred to me that the protesters might think I was a mole taking pictures that could incriminate them. I just wanted to understand what was happening.

Gunshots rang out, and a thunderhead of tear gas billowed up between the Justice Department and the Museum of Natural History. The protesters, breaking ranks, pushed out onto the side streets. I ducked my camera and ran. A fog of tear gas drifted around the Reflecting Pool. It was like a nightmare, worse than death. I could imagine Nazi storm troopers mindlessly goose-stepping through barren trees. I turned and searched for Maggi. She was gone, swept along into one of the other streams of protesters pouring up the side streets.

I ran past the Commerce Department looking for her, and as I crossed the Ellipse, police mustered on the edge of the Mall near the Washington Monument. Gas masks in place, visors down, helmets on, shields up, nightsticks in hand, pistols in holsters, they looked as fearsome as storm troopers. Nothing scared me more than well-armed men mindlessly carrying out the orders of abstract authority. As soon as the American flag was torn down from the pole outside the Justice Department, I knew the police wouldn't wait any longer. The least I could do was take their picture, record their awesome aspect. The Black Panther flag went up the Justice Department pole to a chorus of cheers.

As I started down the row snapping pictures of policemen, a big sergeant yelled out, "Here's our first prisoner, men!" and grabbed me by the arm before I knew what was happening.

Even in my most paranoid distortions of reality, I had not expected to be arrested. Although I offered no resistance, the sergeant shook me, as if to say, "What the hell do you think you're doing, resisting our authority?"

He led me to a black metal paddy wagon and pushed me inside. It was empty, with wooden benches and heavy wire screens over the windows. As the police started their sweep across the Mall, small campfires flickered on the grass and ragged protesters stood around, looking like soldiers after the battle of Antietam. The police fired tear gas canisters and advanced in formation. Every time they threw someone new in the paddy wagon, a cloud of tear gas wafted in. I choked, coughed, and held a handkerchief over my nose. In no time, they had the paddy wagon full of protesters, all choking on tear gas.

"If you have drugs, throw them on the floor now!" one of the protesters said, coughing into his handkerchief.

"I haven't got drugs," I said, batting at the gas with my hand.

"Good, because that's what they're really after." He buried his face in his handkerchief.

The paddy wagon crossed the Mall, then turned east on Madison, north on Fourteenth Street, to the jail in the basement of the Municipal Building. While the outside was like a marble Masonic temple, the jail in the basement was a damp, windowless concrete-block dungeon with chipped and cracked paint. We were lined up before the rotund desk sergeant laboring over a stack of forms.

"Profession?" the desk sergeant was saying as he worked the sheets.

"Geophysicist," I replied.

"Geo—what?" He scowled at me.

"I study the physics of the earth."

"I don't care what the fuck you study," he said through layers of double chins. "Geo . . . Here, you write it."

The rest of the questions were routine. Address, why arrested—he wrote them with the sacramental deliberateness of a monk perpetually copying illuminated scripture. He confiscated my camera bag, gave me a receipt, and led a dozen of us into the cellblock.

The windowless concrete-block cells with single open toilets were overflowing, ten men in a cell built for four: wild-eyed revolutionaries, free radicals of the New Left, homeless ruffians, Weather Underground, jaded insurgents, culprits of sedition, and experienced dissidents. All the American heroes were incarcerated: Uncle Sam, Mighty Mouse, Sitting Bull, Paul Revere, Mark Twain, Hoss Cartwright, and Ben Franklin. They were whooping it up, having a party, whistling, and letting out catcalls and Bronx cheers. Every time the turnkey walked by, choruses of "pig, pig, pig," and "oink, oink, oink," rang out. The turnkey looked straight ahead, like a stripper avoiding eye contact with jeering drunks.

"Hey, turnkey, let me out, I gotta take a dump!" yelled Hoss Cartwright.

"There's a toilet in the cell; use that," growled the turnkey.

"In front of all these weirdos? No way."

"Suit yourself."

"Okay, okay, everybody look the other way." He pulled down his pants, sat on the cracked and stained toilet, covered up with his ten-gallon hat, and let out a raucous fart.

"Oh, God, that's worse than tear gas. Turnkey, we need gas masks!" yelled out Uncle Sam. There was general laughter all around.

At first, I felt no shame at having been thrown in jail; in fact, I was proud and righteous about it. I could brag about it to Maggi. One of my boyhood heroes, Henry David Thoreau, advocated arrest as part of civil disobedience, and he himself was jailed in 1846 for refusing to pay taxes that would support the war on Mexico. As the story goes, his friend Ralph Waldo Emerson visited him in jail and said, "Why, Henry, what are you doing in there?"

Thoreau supposedly answered, "What are you doing out there?"

I felt the same way. It was good to be in jail. It was the people who weren't in jail who were wrong. I had not trespassed or failed to move on when told—they hadn't told me anything. I took pictures of policemen while walking along the street. I felt camaraderie and even revelry with my fellow inmates.

"The ACLU will have us out in no time," Uncle Sam informed me.

Another cell mate, a man dressed like an insurance salesman, in brown suit and red necktie, hung on the bars with his back to us.

"I'm not one of these people," he yelled at the turnkey. "You made a mistake!" He pounded on the bars.

"Yeah, right on!" yelled a hippie in a red headband and leather vest. "I'm not one of these people either. Let me go!" He shook the bars next to the man in the brown suit. A general round of laughter and jeers from all the cells.

"Yeah, man, none of us is one of these people. Let us all go!" Ben Franklin yelled out.

"I'm an American; these people are communists," the insurance man yelled to the metal door as it slammed behind the turnkey. "You can't do this to me!"

"We're all Americans, actually," the hippie yelled at the closed metal door.

More general laughter. Fifteen minutes later the turnkey returned with a new group of inmates.

The insurance salesman pleaded again, this time pitifully. "I'm innocent; you can't do this to me."

The turnkey walked back, ignoring him.

"Hey, man," said the hippie in the red headband, "we're all Americans and they're doing it to us. What makes you special?"

"I wasn't protesting!" said the businessman with exasperation in his voice. "I was just walking down the street!"

"We know, man," said the hippie compassionately. "That was all a lot of us were doing. In fact, some of the people in here are doctors who were trying to help the injured."

The insurance salesman slid to the floor and slumped over, knuckles bruised and bleeding from pounding on the bars.

"You can't do this to me. . . . I'm an American," he said, and dropped his head.

"Yeah, sure," said the hippie, sliding down the bars and sitting on the floor next to him.

"My boss will fire me if he finds out." The insurance salesman turned his back on the hippie.

"Hey, man, it's okay; at least you got a job." The hippie patted him on the shoulder.

"My family will disown me," the man sobbed.

"No, they won't."

Shortly the turnkey led in an ACLU lawyer dressed in a black suit with a red tie.

"ACLU, good; we're out of here," cried Uncle Sam.

"Great!" I said.

"Now, listen up, everybody," yelled the ACLU lawyer. The protesters pushed against the bars. "We've posted bail on the trespassing charges. When I read your name, signal the sergeant and he'll let you out." He started reading names from a list on a clipboard.

One after the other, they released the protesters in our cell, but they didn't call my name. Me, the hippie, and the insurance salesman remained. What's going on? I thought to myself. Now that everybody else was leaving, I wanted to leave too. With the jingle of keys and the clank of cell doors, the turnkey opened and closed the cells one after the other.

The lawyer had read off a full three hundred and twenty names. Finally the insurance salesman was released. He chased the turnkey toward the metal door. "About time; I'm going to sue you for wrongful prosecution."

"You haven't been prosecuted, brother," yelled the hippie after him.

The insurance salesman turned to the hippie. "What about you, aren't you coming?"

"No, I don't think I'll be coming," replied the hippie.

The insurance salesman turned, stepped up to the cells, shook the hippie's hand through the bars, and gave him his card. "Look me up, would you? Seriously, look me up." He smiled.

Only the hippie in the red headband and myself remained. The clang of the cell doors reverberated louder as each tenant left. An alarming silence and a terrible feeling of loneliness filled the cellblock. I felt rejected and left behind. They're never going to let me go, I thought, because I want to leave too bad. I've got to stop wanting to leave if I really want to leave.

"Hey, there's some mistake; you didn't read my name," I said anxiously to the lawyer.

"What's your name?" asked the lawyer.

"Hart, Roger Hart."

The lawyer scanned his list and announced, "They got you on possession of narcotics."

"Narcotics? I don't have any narcotics?" I protested.

"They found a loose Valium in your camera bag," he replied.

"Prescription drugs, I have a prescription for those," I said in despair.

"Sorry, but the pills are not in their bottles. Do you have the prescription bottle?" he asked.

"Yes, but not with me," I replied.

"None of the bondsmen will touch you. I'm sorry." He shook

his head slowly. "You'll get a phone call. You got somebody in D.C.?' "

"Yeah, a girl . . . a friend."

"Have her bring the prescriptions to our office on Monday," he said. "Here's my card."

"But I didn't do anything," I protested. "I'm innocent!"

"Sorry, that's the way it goes." The lawyer shrugged his shoulders, turned on his heels, and walked out.

Oh, shit! Two nights at least; that's worse than I expected.

I was foolish to have hoped for release, to have stopped expecting the worst. But, as in this case, sometimes the worst is beyond imagining. In jail, you can't expect the best or you'll go insane; hope only adds to the intrinsic frustration. Two nights! Thoreau had spent barely one night. For one night you can keep hope; two nights was frustration, especially when your comrades have left you behind. It saps the courage to think you are suffering alone. I would have felt better if they had told me there was no chance of release right from the beginning.

I turned to the hippie in the red headband, now sitting on a wooden bed platform. He looked like one of the Three Musketeers, with long brown hair, pointed black beard, and love beads around his neck.

"Shit, man, I thought I'd be out by now," I said.

"Drugs." He shook his head. "That's what they're after."

"I didn't know I had drugs," I said. "Some Valium must have spilled in my camera case. God, I wish I had one now. What about you? What'd they get you for?"

"Decorating public property." He smiled sheepishly.

"Decorating public property?" I sat down next to him.

"They said it was defacing—a purple peace symbol on the Washington Monument," he replied. "Decorating—defacing, it's a matter of taste."

"Big deal, they'll just wash it off," I said.

"No, they're sandblasting the whole monument. My bail is fifteen grand. No big deal." He smiled broadly.

"Your bail is fifteen thousand dollars?" I was amazed

"Well, my father—I told the pigs my father is a rich man. Big mistake." He shook his head.

I was getting to like his attitude, dignity, and humor.

"What's your name, brother?" he asked.

"Roger, you?"

"Jeremy, brother." He extended his hand. When I took it in a traditional handshake, he twisted my hand sideways and grasped me around the thumb. I did the same to his hand.

"What about you?" he asked. "Why did they arrest you?"

"I was taking photographs," I replied. "I'd never been to a demonstration before. I'm here with a friend."

"Take out the press first, no photographs. They always do that."

"Phone call." The turnkey opened up the cell door without looking at us.

"Follow me." He led me down the hall to a pay phone. I dialed the hotel and they put me through to Maggi's room.

"Maggi," I said into the receiver.

"Roger, is that you?" she replied.

"Yeah."

"I've been worried. Where are you?" she asked.

"In jail. I got arrested taking pictures of the police."

"Far out, fantastic!" she said. "I mean . . . Oh, my God. When are you going to get out?"

"I don't know," I replied. "They found Valium."

"But they're prescription," she replied.

"They weren't in the bottle," I said with exasperation.

"Ohhh," she replied.

"Maggi," I said, "can you get the bottle out of my pack and bring it to the ACLU on Monday?" I read the address to her.

"Of course. Oh, God, you're in jail!" she exclaimed.

"There's just two of us left; everybody else has been released," I said glumly.

"How you doing?" she asked.

"Not well. It was okay until everybody else left. Now it's depressing . . . being left behind. One part of my mind is saying, no big deal, you got arrested for taking pictures of policemen. On the other hand, just being in jail, the way the police treat me, makes me feel like a criminal. I feel like I've already been convicted and punished. Shame is the worst punishment. Perhaps I just need some Valium, who knows?"

"Don't worry," she said. "You'll be okay. I'll get you out on Monday, and, Roger, I love you."

"I love you; see you Monday," I replied.

"See you Monday, love." She hung up the phone.

The turnkey led me back to the cell and soon slid two trays through a slot in the barred door, bologna sandwiches, on dry Wonder bread, no lettuce or mayo.

Jeremy tasted his Wonder bread sandwich and gagged twice. "Dry as a hair ball," he said to the turnkey. "Brother, don't you have any real food? This is fascist food."

"I ain't your goddamn brother," retorted the turnkey. "What do you expect, you're in jail! It's your own goddamn fault. If you don't love this country, why don't you leave it?"

"I do love this country," replied Jeremy. "That's why I want to stop the senseless killing of my brothers."

"It's not senseless, and they're sure as hell not your brothers. Now shut up or I'll cut off your Wonder bread."

"Listen, brother, have you ever dropped acid?" asked Jeremy.

"What do you think?" snorted the guard, curling his lip.

"If you'd ever dropped acid, you'd know we're all brothers. Just by putting my hand on these bars, I can feel everybody who's ever touched them before me." Jeremy stroked the bars.

"That makes you some kind of holy man, doesn't it?" replied the guard. Smiling maniacally, he grabbed Jeremy's love beads through the bars.

"No. I just wish you peace, brother." Jeremy struggled and pushed against the bars with the palms of his hands,

"Let me tell you something, holy man," said the turnkey, his face suddenly red, fierce, and menacing. "If you were my brother, I'd shoot you myself." He let go of Jeremy, turned his back, and walked off slamming the metal door behind him.

"Oh, wow! Wow. Oh, bummer, man; that turnkey's a real turkey—so to speak." Jeremy shook his head, sat down on the bed platform, and turned to me. "You ever drop acid?"

"Not that I know of."

"It's like you make contact with this part of your mind that is in contact with a collective source that everybody has access to—well, everybody except the Wonder bread turnkey."

We fell into a silence opaque with thought. Jeremy turned to me. "Did you make it to the Amazon?"

"Sure, I worked there."

"The Amazon, man . . . what's the Amazon like?"

"Surprisingly, it's not a jungle, except along the riverbanks.

Hardly anything grows on the ground under the forest canopy, and there's hardly any soil. Everything rots right away and is sucked up by the trees," I replied.

"What were you doing there?" He stood up and walked to the cell door and looked out through the bars at the metal door where the turnkey had disappeared.

"I was a geophysicist for an oil company. We found oil, lots of it."

"Why'd you leave?"

"I couldn't stand to see the petroleum companies taking over Indian land . . . they firebombed whole villages with women and children," I said. "Now my passport has been taken away and I can't go back."

"Why, man? I thought passports were our personal property as citizens," said Jeremy.

"I did too, but I was on a mountain-climbing expedition that crossed two miles into Tibet. I didn't fully appreciate the situation at the time, but technically, since Tibet was taken over by the Chinese communists, I illegally traveled in a communist country that was restricted on my passport. Then, I was on another mountain-climbing expedition to Chile and we all got kicked out as spies."

"Far out, man . . . a spy?"

"Well . . . the Chilean government thought we were spies . . . partly because we had an Argentinean—the Chileans hate the Argentineans—with us and partly because the other members went to an area without the permission of the minister of defense."

"You're a fuckin' spy, man?"

"Yes, sir, I'm in the book at the Antofagasta border," I boasted. "I'll tell you, I'm beginning to feel like a spy. Like my friend, Peter, he wasn't a spy, but once the Argentineans heard that he was kicked out of Chile as a spy, they picked him up and interrogated him. They, in essence, turned him into a spy. My congressmen and senators, they all tell me I broke the rules; I got to pay the price. I don't even know what the fucking rules are. Man, I can't seem to do anything right. I was just taking photographs, and now this. Oh, shit! Man, my father's going to have a conniption." I took a bite of the bologna sandwich, and slid back on the pallet.

"Hey, don't feel bad, man . . . 'Scuse me, I gotta whiz," said Jeremy. "There's a lot of us in the same boat. As a famous man once said, 'Under a government that imprisons any unjustly, the true place for a just man is also prison." He pissed into the coverless brown-stained toilet.

"That's good; I'll tell my father. Who said it?" I asked.

"It was Thoreau." He zipped up and kicked the flushing handle with his foot.

"Thoreau?" I yelled above the sound of the flushing toilet.

"Thoreau said the jail thing."

The turnkey waddled down the row. "Five minutes to lights out."

I rolled up my coat and stuck it under my head as a pillow.

I awoke, depressed and anxious, with a terrible headache and a heartache to go with it. Was I staying in jail or getting out? The uncertainty was nerve-racking. I was ready to suck up to anybody who promised to release me. It was a good thing I hadn't been entrusted with important secrets because I would have spilled the beans in a flash if I thought it would save me from spending one more night in jail.

Jeremy sat up. "Hi, man, morning; sorry; I gotta take a dump bad." He pulled down his pants and sat down on the toilet. "Sorry, man, this one's a stink bomb."

"How did Gandhi do it?" I asked.

"Do what?" He lifted up one cheek and farted.

"Spend all that time in jail and keep his sanity."

"Well, first of all, he was in a cushy jail, secondly he had his wife with him, and thirdly rich people all over the world were reading and publishing his writings. Besides, he probably got to go to the latrine in private."

The turnkey brought breakfast—cornflakes, milk in a small carton, day-old jelly doughnuts, and sour overprocessed coffee in plastic cups.

"This will bounce the old blood sugar," said Jeremy.

"For about twenty minutes," I replied glumly.

"Don't worry, man; everything's all right. Hey, why don't you grow your hair long?" He brushed his long black hair with his hand.

"Why?" I patted down my short curly hair.

"You'd look good like a pioneer or mountain man."

It was Sunday, and a minister in white collar, with flowing white hair and crystal clear bifocals, showed up.

"Mind if I talk to you for a while?" He clutched a Bible to his breast.

We were getting bored, so we agreed to talk to him. The turnkey let him into our cell.

"What are you fellows in jail for this fine Sunday morning?" He smiled beatifically.

"We were arrested in the demonstration," said Jeremy.

"The antiwar demonstration?" asked the minister. "Mind if I sit down?"

"Sure. No, the peace demonstration," said Jeremy.

"Are you boys against the war?" He sat down, placed his Bible on his lap, and crossed his pasty white hands over it.

"He is." I pointed to Jeremy.

The minister turned to Jeremy. "Are you against defending Christianity on foreign soil?"

"Absolutely. They're Buddhists. What right do we have to bomb them into submission?" Jeremy shook his hair and let his arms dangle loosely between his legs. "Besides, didn't Christ say to turn the other cheek?"

"Well, I hardly think we are bombing them into submission, and we are not supposed to take the Bible literally." He looked at me. "May I ask what is your religion?"

"I was brought up Methodist," I said. "Now I'm a transcendentalist."

"A what?"

"I find spiritual fulfillment in nature, kind of like a Druid."

"Paganism was discredited centuries ago," said the minister, turning red at the tips of his ears. "You are an anachronism, moving backward. Christ is for those of us seeking progress. Don't you want to be part of progress?"

"Why must destroying nature be part of progress?" I asked.

"God made nature for our use," he said. "Besides, nobody else was using it."

"I was using it . . ."

"We can't hold back progr—"

". . . me and about twenty or thirty million other pagans, you know Native Americans, North and South!" I was starting to get pissed off.

"Well, you can't blame religion for destroying their cultures."

"Oh, yes, I can. I saw dozens of cultures destroyed in the

Amazon. The missionaries made contact with the Indians, converted them, softened them up, and then the oil companies and military came in and took the land away. Your self-righteous attitude is the foundation of genocide and militarism the world over."

"Wow, man." Jeremy looked at me in disbelief.

"It's true," I said softly to him.

"Now, hold on!" The minister's face turned red. "We are trying to help these people; we watch after them."

"Yeah, then how come Christian churches are the biggest property owners on earth. What do they do with all that wealth?" Jeremy chimed in.

"It's true that some churches have accrued wealth." The minister's face shook as it simmered in his rage. "But not all of them."

"Is it right for you guys to cut down the forests and dig up the minerals of the earth to fashion your altars from?" I was getting into it, feeling strong in my rage. I didn't know how Jeremy remained so calm.

"Let me ask you a question," said Jeremy.

"Fire away," said the minister.

"Have you ever been high on acid?"

"What do you think?"

"I guess not."

"That's right; I get high on Christ."

"If you really got high, you would see that we are all brothers."

"I can see you really don't want to discuss this in a rational manner; you will never find Jesus Christ with that attitude."

"What makes you the authority on who will and who won't find Christ?" asked Jeremy.

"The Church, that's what."

"You can take your church and shove it."

"Humph. Why don't you become Muslims like all the black athletes?"

"Because I can't stand blind obedience to authority of any kind," Jeremy replied.

"It's that attitude that got you in jail. Guard! You're always going to be in trouble with that attitude. You think too much. An open mind is a hole in the head. I give you one last chance. Do you accept Jesus Christ as your personal savior on this day of our Lord?" He stood up and held his arms above Jeremy.

"Yes, I do, and Buddha, Krishna, and Mohammed." Jeremy stood up and raised his arms above the minister.

They were standing there, arms raised above each other, when the turnkey opened the cell door. "Everything okay here?"

"Just fine," said the minister out of the side of his mouth. "Well, I do hope you will change your mind." He dropped his arms with a disgusted look on his face and stepped out of the cell. "If you need any help, call me." He handed me a card. "I'll include you in my prayers. Have a good day." He turned and walked away.

"Peace, brother," said Jeremy, "and I'll include you in *my* prayers."

The minister left. I looked at Jeremy quizzically.

"What?" He held his palms up. "Of course, I'll include him in my prayers."

We had lunch of Wonder bread and bologna—same menu as the day before, perhaps the same sandwiches.

"Man does not live on Wonder bread alone," muttered Jeremy.

"Hey! No more blasphemy," I joked

The next morning the turnkey came to let Jeremy out before he even had breakfast. Jeremy turned to me, took his beads off, and placed them over my head. "Hey, they make you look like a warrior. Peace, brother." He hugged me.

Shortly after lunch it was my turn to be released. Maggi was in the waiting room, and when she saw me, rushed up and threw her arms around me. I was still depressed and weakly hugged her back.

"Oh, God, I need a Valium," I said.

Maggi held the bottle up in front of me. "Here you go." She took me by the arm and led me up Sixth Street. "We can catch a bus from there to the State Department," she said. "You still want to go there, don't you?"

"I don't know what I want to do," I said. "I feel weird. Well, what the hell, we're here; things couldn't get any worse. Let's walk there."

I'd expected that Washington, D.C., our nation's capital, would be more impressive. The lifeless avenues, devoid of friendliness—there was not even a billboard, public rest room, or bit of graffiti—were built to instill awe and respect in the institutions of government without a bit of wit or turn of architectural grace. The streets and lawns, isolated from nature by their rigid geometry, conveyed respect for linear authority. The jail looked the same as the Justice

Department, the State Department, the hospitals, the airports. Separate institutions were combined into one mega-institution.

The place made me apathetic, crushed my mood, and made me feel as if I absolutely didn't matter. I'd thought once I got out of jail, I would feel better. Instead, I felt worse, intimidated, threatened, and, for the first time, realized my government didn't give a damn about me. The last thing it was going to do was help me.

I showed a clerk at the front desk of the State Department my passport with the word *VOID* punched in it.

"This is a mistake," I said.

"Here let me have it," he replied. "I'll show it to the undersecretary."

Time crawled to a stop in the State Department waiting room. A thin, balding man in a three-piece suit came out to the lobby.

"Mr. Hart?" The officious tone of his voice irritated me immediately.

"Yeah. That's me," I replied defensively.

He stood before me with a stiff military posture; shields and auras of arrogance and indifference surrounded him. "I am here to tell you the United States government is revoking your passport privileges."

I had expected this and it was happening anyway.

"I am a U.S. citizen and that passport is my property. We made a mistake. I am sorry," I replied.

He continued in a monotonic voice, "You have violated your privilege by trespassing in restricted areas in violation of the immigration laws of those countries."

"Then isn't it up to those countries to reprimand me?" I asked." Why is it within the jurisdiction of the State Department?"

"We are required to protect our citizens traveling abroad and to maintain the integrity of visa privileges with foreign countries."

"That passport is my personal property," I said.

"No, your passport is property of the U.S. government," he said mechanically.

"I have one question for you," I said in desperation.

"What's that?" he asked.

"Have you ever dropped acid?"

"What do you think?" he snarled.

"Forget it. That passport is *my* property," I retorted.

"You are wrong," he replied. "It is property of the U.S. government. If you don't love this country, why don't you leave it?" His jaw was set firm.

"I'd love to. Give me my passport back."

"Security!" he yelled to the guard near the door.

"That went well," I said to Maggi as we walked down the steps of the State Department.

"He's wrong; you'll just have to hire a lawyer," she said.

As the Greyhound bus sped north over the dark plains of New Jersey, I stared at the acres of oil refineries with flaming torches on their distillation stacks. Maggi dozed with her head on my shoulder. I hated that mess. The consumer reality, fueled by petroleum, was projected to every tribe and culture over the airwaves and in movies. Once indigenous people accepted the consumer culture, they left the land open to exploitation. With advertising, artificial value was created for anything. The advertising juggernaut portrayed nature as dangerous, backward, and alien. Once nature was demonized, the land was taken from the people who lived on it. Once removed from the land, we would lose our traditional reference points and abstract authority could be hyped as the new reality.

I didn't understand New Jersey, or the rest of America, for that matter. My brushes with death on Mount Everest and in Tierra del Fuego had introduced me to a new experience that made the constant striving for material success and recognition seem meaningless. In fact, I found it difficult to talk to others preoccupied with self, time, moral judgments, and utilitarian considerations. I became ashamed of acquaintances who were cocksure and preoccupied with self-assertion and overvalued words.

My experiences had uniquely affected me. I felt isolated, lonely, unsure of myself. Unable to speak to my friends and family, I was unable to look them in the eye but, at the same time, intimidated by the wealth, cars, and houses they and their children began to accumulate.

Through a simple calculation, I had realized that the consumer abundance promised by propaganda could not be provided to everybody on earth. The petroleum resource is limited. For one unique period in earth's history, Cretaceous plankton flourished in an equatorial sea. Their dead bodies accumulated in great quantities on the bottom of the sea and were rapidly blanketed with layers of sediment that turned into rock strata over eons. The plankton decomposed into petroleum, and the consumer culture was fueled by burning their decomposed remains. But there were not enough resources to sustain the promised lifestyle to every man, woman, and child on earth. We were told that nuclear fusion would supply unlimited energy. But that was just more propaganda. Fusion requires enormous pressures to contain the reaction at temperatures that would melt any known containment vessel on earth. No, to me, the promise of unlimited energy was just more propaganda to sell research into nuclear weapons.

Given that there were not enough resources to sustain the consumer dream for the whole human population, only the affluent could afford it. Thus it was inherently elitist. Yet it had no intrinsic value, no capacity to bring happiness, its only value coming from the envy the rich can instill in the have-nots. Even the ad executives believed their own hype, and most people in the world believed that with more education, more skills, more attitude, they, too, could reach the promised land of consumer fulfillment.

I didn't believe the hype. The land of consumer fulfillment was an illusion that always left me unsatisfied. The garages of suburbia were full of trinkets, tried once or twice, then ignored. Weed Eaters, hedge trimmers, fertilizers, pruning saws, rotary mowers, rototillers, barbecues, hibachis, piled up on the patios of suburbia, while Quechua Indians in Peru still plowed their potato fields with foot trowels fashioned out of old car springs and a tree branch. Millions of poor farmers around the world would give five years, income for the simplest rototiller discarded in the hobby gardens of America. I had seen both realities, and having traveled in the poor countries, I could no longer accept the illusion of consumer greed. Maggi, because she had lived in poverty in Peru, understood my point of view.

Worse yet, like in my dream of the Incan priest, the power of nature was being washed from the face of the earth to be replaced with a sense of desperation that, in my opinion, would lead men to deny acts of empathy.

As the flames and fumes licked and curled upward and joined the mega–smog bank hovering over the East Coast, I despaired. And I felt even worse because I had let New Jersey depress me. Hey, if this is inevitable, deal with it, get on with your life. My brain was scrambled. Everything I had been taught, everything I believed in, seemed open to question. I had no ideals left to hold on to. I became sullen in my own country, embarrassed by my countrymen, uncertain of how to relate to them, and longed for the day when I could leave again. I had only the tentative comfort of Maggi, her calm peace with the world. She was my only hope and my only concern now. At any moment, I could lose that too.

Maggi lifted her head. "Where are we?" She looked out the window.

"New Jersey," I replied.

"I'm going to the john." She squeezed by me and made her way to the rear of the bus.

Returning to her seat, she stumbled and fell on my lap. I put my arms around her and hugged her. She kissed me and whispered in my ear, "I love you."

"I love you so much. Please stay with me tonight," I replied.

"I promised Roger Whitehouse I wouldn't."

"Please." I didn't mind sharing Maggi, but I couldn't stand the idea of being without her.

"It's just too difficult." She pushed me back and looked at me.

"Please." I kissed her again and caressed her breast.

She took a deep breath and sighed. "It's so hard to say no. Okay."

That night we stayed together at the Times Square Hotel. The next day I took a bus to Boston for a visit with my parents, whom I hadn't seen in over three years.

CHAPTER SIX

PRECOGNITION

In the grand view of events, being arrested in a peace demonstration or having a passport voided were not major violations of morality; nonetheless, my parents were bound to take the news badly. I dreaded the impending encounter with them. At some level I felt like a criminal, and I knew a further rift could develop within my family. Throughout my youth I had found solace in the woods near my father's house, and I visited those woods again on May 28, 1969, in an attempt to compose myself for the forthcoming confrontation. I found that, like many things, the woods had changed.

I hopped over boulders on the way down to Walden Pond, not Thoreau's Walden but the one in Lynn Woods, north of Boston. The staccato cry of a flicker startled me as it darted between the black oaks. I turned onto the trail my mother and I followed when we picked blueberries. The lustrous back of a black racer snake slipped into the dark shadows of a crack in the white rock. The hot sun reflecting on the granite ledges kicked up the smell of sweet fern and memories of my mother picking blueberries on summer days.

As I approached Walden Pond, uneasiness came over me. The harmony of the woods was compromised by trash and ragged furniture lining the roads. I stumbled across a bag of pornographic pictures. The fringe where society meets nature is the strangest of all, the worst of both worlds. Weeds and damp leaves festered in mold and decomposition among fragments of trash. The woods cringed under the pall of pollution that hovered overhead. Acid rain wilted and stunted the black oaks as if they had lost heart and stopped growing. Countless forest fires, set off by careless cigarettes, had consumed and scorched the soil, turning it black and barren.

My grandfathers had worked within the designs of nature. My mother's father was a chicken farmer; my father's father was a wood-carver, stonemason, and ice cutter in Gloucester. Our neurochemistry had coevolved with nature. Now, within one generation, technology was destroying the process that had founded our own biochemistry. I felt like an anachronism, unwilling and seemingly unable to take the step into technocracy and totalitarianism. I was frightened of the prospect of having to give up my worldview and despaired even more at my own silly petulance. The earth is trashed, the seas are polluted, the skies are somber, and the weather patterns pushed into unprecedented directions like a wild beast. Get over it, and get on with your life! I told myself.

My father, a genius out of poverty, had gotten scholarships to MIT. In less than three months we were about to land the first men on the moon, and the eyes of the world were focused on that event with the hope that space was a new frontier that offered resources, progress, and hope for the future expansion of humankind. Some spoke of it as a dumping ground for nuclear waste. Space flight was the ultimate victory of linear physics—Galileo's trajectories, Kepler's orbits, and Newton's laws of motion. Unlike space flight, which we could control and regulate, the forest was wild, complex, and unpredictable. The initiative into space made the trashing of Earth easier. We were no longer dependent on nature and could always move to the Moon or Mars if we soiled our own nest on Earth. That seemed to be the attitude; little did we know at the time that the Moon, Mars, or any other planet in the solar system is grossly uninhabitable.

I walked along the shore of Walden Pond. This artificial, man-dammed lake stank of pond scum stranded on the shoreline after the water level had been reduced by the needs of the water-using public; boulders and sunken stumps populated the beachless shore.

I came upon the wolf pitfalls, stone-walled traps eight feet deep. Once considered a threat to the farming lifestyle for preying on chickens, sheep, and other livestock, wolves had long ago been obliterated from the East Coast. Now that the wolves are gone, I thought to myself, the woods are safe for people to dump their trash.

I followed a trail to the top of one of the bare granite domes, where I got a view of the skyline of Boston to the south.

I remembered that I had been here in the middle of winter a decade ago—snowdrifts sprinkled with bird footprints, chickadees, grosbeaks, nuthatches, and waxwings feeding on seeds in the pinecones. Oblivious to my presence, preoccupied with their feasting, they produced a cacophony of cracking, chirping, and flitting about that made me feel as if I were floating over the earth.

Now, I saw the rusty remains of an airplane that had crashed in the bushes. The thought of the attendant horror, smoke, and blood evoked visions of sheer terror that stretched over the landscape, wiping out any memory of ecstasy, any memory of the bird feast. Even in the peace of the woods, disasters of technology fell from the sky. Perhaps I was flipping out from too much Valium. It didn't matter.

I stood on the edge of the cliff and looked out over the sea of trees rippling to the skyline of Boston. The wind blew across the leaves in wave after wave. Perhaps to get back at my parents for some reason, I was purposely breaking my life into shards that I could never put back together. But how could I have planned to lose my passport or get arrested? No, there were other forces at play.

The trees below my feet rustled and billowed in the wind like parachute silk. The canopy would catch me, let me down softly. I had lost everything I could hold on to. The emotional descent I was suffering pulled like an undertow. The first raindrops sizzled on the hot granite and kicked up the smell of seared earth. What was the point of struggle? What was the point of staying alive if death was so good? By defocusing my eyes, I could make the trees below come up and go back. I could let go, so easily. A crow hurried over the tops of the trees. The wind picked up and rocked me. I closed my eyes, shut them tight. But what if I didn't die? What if I disappeared into the canopy of trees, unable to find my way back? The crow called out; the wind pushed me backward; I opened my eyes, threw up my arms, and fell back onto a ledge. I scrambled back to safety on the top of the cliff and sat down, catching my breath and my wits.

The end of the blueberry trail emerged into a cleared area under the power line towers that marched in a straight line over the cliffs. High-gauge wires swung over the landscape, buzzing with ionizing radiation; the bushes underneath curled and twisted in frenzies of mutation. I followed the trail back toward my father's house.

Playing my Peruvian bamboo flute helped me to feel better, as if the music cut a swath through my anxiety about facing my father with the news of my arrest and loss of passport.

A wall of freshly blasted granite boulders blocked the trail. I climbed over them and emerged onto a newly planted lawn that smelled of Vigoro fertilizers and stretched in a continuous field behind a row of five houses. Concrete-block basement walls supported the clapboard houses, each with an identical open deck and wooden steps leading down to the monotonous green of lawn. Clothes hung drying on aluminum poles, green hoses were coiled on plastic racks, and the lawns were dotted with furniture, inflatable swimming pools, and flotation toys.

When I stepped onto the soft lawn and headed for the passage between houses, a husky potbellied man emerged onto the porch holding a can of beer.

"Hey, you!" he shouted at me. "What are you doing here?"

"This is the way home." I pointed past the house to the pavement. "I live down the street."

"What house?" He took a swig of beer.

"The Harts. I've been away for a while." I took a few tentative steps onto the lawn.

"Hey! You're wrecking my new lawn."

"No, I'm not. It springs right back; look!" I gingerly stepped forward.

"This is my yard!" he said angrily. "I don't want you crossing here."

"The path comes out of the woods right there." I turned, backed up a step, and pointed to the faint path leading through the trees from the wall of boulders. It was a good mile retreat to the beginning of the loop. "My mother and I have been using that path for twenty-five years."

Then he saw the beads that Jeremy had given me. "What are you, some kind of fuckin' hippie?"

"I won't do it again," I muttered. "Look no footprints even." I took a step forward.

"You won't do it now! Go back or I'll call the cops." He set the beer down on the handrail of the porch and shook his fist at me.

"It's just a few steps out onto the street; I'll be careful." I scampered out to the street.

"Goddamn you! I'm calling the police."

As I walked up the asphalt driveway to my father's house, the flag hung limp on its pole. The white oaks—well pruned, healthy, and in full green foliage—framed the white clapboard house with its set of three dormer windows. When I closed the door behind me, the knocker clanged on the outside.

"Roger, is that you?" yelled my mother urgently from the kitchen.

"Yeah," I replied.

"Come in here, I have something to ask you."

"Okay." I walked to the kitchen.

"Here, sit down." She pulled a chair out and slid it to the middle of the linoleum floor.

I stared at the chair for a few seconds. "No thanks, I'll stand."

"Dick McDermott up the street called," she said with a nervous laugh. "He said you walked across his new lawn."

"It blocks the old blueberry path. The one into the woods," I replied.

"He doesn't want you using it anymore." She frowned at me.

Does he think he owns Lynn Woods? I thought to myself and said, "I didn't hurt his lawn."

"I'm sure not," she replied. "But he doesn't want you walking on his newly seeded grass."

"He's just being hyper—"

"And he said you were rude to him," she interrupted.

"I didn't mean to be." I turned to our back porch, a large screened enclosure that was the family's retreat on warm summer nights.

My father was watching the six o'clock news. He set down his pipe, reached forward, and turned off the TV when I sat down on the wicker couch.

"So, you've been in Washington?" He turned to me, stiff with arthritis.

"Yeah, a peace demonstration," I replied.

"What about these demonstrations? They tried to blockade my lab at Harvard." My father gestured at me with the stem of his pipe. As an electronics engineer, he had helped develop radar and supervised the construction of sonar nets across harbor entrances in the South Pacific during the war.

"They probably thought you're working on some kind of military project," I said.

"I'm not." He knocked the pipe against an ashtray, scooped out the ashes with a pipe cleaner, and blew through the clean stem. "I don't see what they have against me. My inventions help people."

"That's true; they just feel threatened by technology. They think the war is wrong," I said.

"My inventions aren't involved in the war," he replied.

"Some technology *is* destroying the natural world."

"No," replied my father. "Technology is natural, the survival of the fittest. Nature showed us that."

"Yes, but the fittest are not always the killers. Some species survive through cooperation rather than competition," I replied.

"You just say that because you're not a competitor. Have you ever won a competition?" he asked. "We won the war by kicking butt—there's no greater success."

I can't believe I'm hearing this, I said to myself. My mother carried out a pot of blueberries she had picked in the garden and sat next to me. The aroma of the berries was slightly tart, like the smell of the wild ones we had picked in my youth.

In the hot summers before I entered grammar school, my mother and I, pails in hand, scouted through the underbrush of Lynn Woods for good patches of blueberries. Kneeling, we rolled the ripe berries into our cupped hands with our thumbs, like milking a cow. At first, I stripped green berries along with the blue ones. By the third summer, my pail filled almost as fast as my mother's with berries almost as blue as hers. Even though I was barely tall enough to see over the bushes and ferns, I learned to wander off on my own, to hunt for my own berry patches, using the black oaks and granite ledges as route markers. Under my mother's watchful presence, I developed calm assurance that it was safe to be lulled under the rustling leaves of the oak trees. I stumbled over ledges, thrashed through bushes, watched the flight of a bird, and pulled up short at an encounter with a skunk or porcupine. My brain-cell dendrites branched with the brown woods paths that led from patch to patch of blueberries. In Lynn Woods, near the shores of Walden Pond, among the cliffs, trees, and ferns, the smell of wintergreen plants, the pink lady's slippers, I put together my worldview while picking blueberries with my mother.

At night in bed, in the drowsy minutes before sleep, when my brain was quiet, images of blueberries swam through my mind and floated on the inside of my eyelids. I couldn't tell if the black-light designs were within my brain or projected from an outside source. Sometimes, instead of blueberries, I would see faces, a pleasant slide show, one face after the other, of people I had never known or met.

"It's time you settled down and came home." She positioned the pot of blueberries on her lap.

"I'm not sure I can settle down."

"Why not?" She plucked off a blueberry stem with her fingernails.

"Something happened during the fall on Everest, when I thought I was going to die."

"Oh, dear!" said my mother.

"Once I accepted my fate—once all the little thoughts in my brain stopped, I said to myself, *'This is strange: here you are about to die and you feel wonderful!'* " I'll never be able to settle down until I find that feeling again."

"This mountain climbing is too risky; I can't stand to think of you falling off a mountain and dying," she replied.

"If it makes you feel any better, I've given up mountain climbing." As much as I loved mountain climbing, it seemed trivial compared to the heightened awareness I had experienced. I was more interested in exploring altered realities than climbing for its own sake.

"Mount Everest. That's a good example," my father interrupted. "You failed there, didn't even get close to the top. You screwed up and fell. Now you want to turn it into some great enlightening experience."

"It was," I answered.

"And just because you were lost in a blizzard in Tierra del Fuego doesn't mean you have special insights into life." He scooped his pipe through the yellow tin of Dill's Best and tamped the tobacco down.

"My experiences are mine, and yours are yours. We are living in different realities. I'm not saying your reality is wrong," I replied.

"Well, I'm saying your reality is wrong. It's idle fantasy," he retorted, turning his back stiffly to the wind and me, and striking a match.

"Perhaps I am making something out of nothing," I replied. "But my reality is supported by science."

"Ridiculous." He sucked the flame into the bowl and shook the match out angrily. "You cannot prove any of your views."

"Particle physics has shown there is no elementary building block of matter."

"What's that got to do with your near-death experiences?" He blew the smoke out contemptuously.

"There's a whole other reality beyond materialism," I replied halfheartedly. I could see that it was no use. He would never believe me.

"That's just mystical mumbo-jumbo, the stuff of crackpots, a way to avoid the responsibilities of life," he said.

"Perhaps you're right. But perhaps your reality is an illusion and it's your responsibility to figure that out," I replied, but he wasn't listening. His mind had wandered off like the smoke from his pipe.

"Why don't we just drop an atomic bomb on Vietnam and get it over with." My mother hectically pulled the blueberry stems.

I paused and searched for the right words. "That would kill a lot of women and children in addition to soldiers." I immediately knew I'd made the wrong choice.

"So what?" she said. " We killed a lot of Japanese women and children. You were only four. You wouldn't remember, but it saved American lives by getting the war over sooner."

A wave of emotion swept through me. I decided to try another tack. "Why don't we just let them sort it out. It's a civil war. It's like if the English invaded during our Civil War and fought on the side of the South."

"We should just bomb them." Her mouth was set firm, and the blueberry stems piled on the cement floor like machine-gun shell casings. There was no room for further discussion. I filled with frustration. Until that point I never fully understood my parents. I had been too close to them. Every rational perception told me they were wrong. But emotionally, I was strangled. They cut me off because I didn't agree with their prejudices. I needed their love, but I couldn't take back the experiences that led to the estrangement of my worldview from theirs.

"Why did they arrest you?" my father asked.

"I don't know," I said. "Trespassing, I guess."

"You mean you went where you weren't supposed to go?" My mother's voice was full of reprimand.

"It was public land, the Mall around the Washington Monument, Mom; anybody can go there anytime," I said defensively.

"I mean," she replied sternly. "You must have been doing something wrong."

"Taking pictures of policemen?" I asked. "What's wrong with that?"

"They must have told you not to take pictures," she said. "Or told you not to walk on the grass."

"They never said anything. They arrested me just to hassle me, to teach me a lesson." I threw up my hands in despair.

"And did they?" she asked accusingly.

"Yeah. They taught me a lesson." My reply was full of sarcasm, but she didn't seem to notice.

"And what's this about your passport? They took your passport?" My father took up the line of accusation. The smoke from his pipe climbed a street of spinning vortices and faded into the twilight. When I was a child, he entertained us kids by blowing smoke rings.

"Yeah, because of Mount Everest," I said.

"Why," my mother said. "Is that against the rules?"

"No, but we strayed over the border into China," I replied.

"Oh, my, they're communists, aren't they?" asked my mother.

"That stupid, crazy Sayre." My father sucked on his pipe agitatedly.

"Everybody who climbs Mount Everest crosses the Chinese border; it goes right over the top. You step back to take a picture on the summit and you're in China, or actually Tibet, because the Chinese are in Tibet illegally," I replied.

"You broke the rules; just because others do it doesn't make it right." My father's voice dripped with self-righteousness.

"No! No!" I jumped up. "Here, let me show you." My parents kept a copy of *Four Against Everest* on the coffee table so they could show it off to the neighbors.

"See, we crossed into China or Tibet at this point," I said, pointing to East Rongbuk Glacier on one of the maps. "Not more than two miles over the border. In fact, Sir Edmund Hillary did the same thing in 1952; that's how Woody knew it could be done."

"You must have done something wrong or they wouldn't have taken your passport away," said my mother.

"A passport is supposed to be the property of a citizen," I said. "I didn't break any laws. We were just climbing a mountain."

" 'Property of a citizen'? Don't you think we know you are causing all this trouble to rebel against us?" said my father.

"And what about jobs?—why is it you can't hold a job?" asked my mother.

"My encounters with death put me in touch with an experience of being alive that I won't get going to the office every day." I paced up and down the porch.

"You're just making excuses for being lazy," my father blurted out.

"On the contrary—facing death is more work than you can imagine. Going to the office every day is just a crutch to avoid facing the fear of death."

My father rolled his eyes and looked around him angrily. "Your famous fall on Mount Everest, this blizzard thing in Tierra del Fuego, were absolute failures—you failed and you are trying to cover up your failure by ascribing mystical hocus-pocus to it."

"Perhaps, in part—but it is also possible that I experienced something important. I want to find out what it is." I stopped pacing and faced them.

"It's possible to rationalize anything; you are just avoiding your duties," my father said angrily.

"That's just it—these experiences were beyond rationalization. All thoughts completely stopped. Everything in my life dropped away. Duty to family, duty to job, duty to career—they all became astoundingly irrelevant. I had direct experiences of something more real."

"You are deluding yourself. Listen! I fought in the war, studied electrical engineering, and built this house for you. It's clear that you don't appreciate what I've done for you. I wanted to save you from the suffering I went through as a child. Wait until you have a child of your own. Then you'll understand."

I shook my head. I didn't mean to be ungrateful. I could see that I had genuinely hurt his feelings. I knew we were on a precipitous course of alienation and that a word of contrition from me could salvage our relationship, but I was too wrapped up in defending my worldview. How could they possibly think I had deliberately crossed into Tibet or gotten arrested just to piss them off? If I was reacting to them, seeking attention, validation, I had more power—was more organized—than I ever realized. They were critical because I didn't do things their way. My father drove me crazy like I drove him crazy. They were incapable of giving me love, validation, or support now that I would not accept their worldview.

"Are you saying I can't stay here because I disagree with you?" I asked under my breath, throwing up my hands.

"Home is where the heart is," said my father without looking up.

"Well, you must respect me or you wouldn't have a copy of *Four Against Everest* here." I turned and walked away.

"That's different." My father hissed out pipe smoke.

My mother's face puffed up, flustered. "Oh, dear, sit down; I mean, would you sign a copy of your book for Reverend Drake?" she said.

"Sure, but we didn't do anything wrong." I sat down. "What's on TV?"

"Isn't Dean Martin on now?" asked my father, turning and looking at her.

"Oh, yes, Bob; switch him on." My mother calmed down.

I listened as Dean Martin suavely sang, "That's Amore."

Having grown up stifled in the suburb, I would watch anything, even *The Lawrence Welk Show* and *The Dean Martin Comedy Hour*. I could sit next to my parents if all we did was watch TV. It was a substitute for social interaction, a distraction that kept me entertained. It helped me forget the normal problems of life. It relaxed me. It was an illusion that had an effect as powerful as any reality.

I liked TV, but wasn't sure I should. It devastated my attention span, and made me feel I could become something I wasn't, without making clear what that something was. It had jabbed, poked, twitched, probed, and galvanized my mind into a world of illusions. The surrealistic juxtaposition of crime shows and family shows confused me. The images, the cartoon monsters as well as the Saturday morning heroes, were always beyond my grasp. Ads were indistinguishable from serious drama. Even though I knew it was empty, I was still frustrated by the hype and, as a result, I would never be satisfied with what I did have.

I was eight before I started watching TV. But what about the children who are baby-sat by TV rather than their parents? With the laugh track, children hear people laughing at things that are not funny. With the applause signs, things are applauded that are not laudable—infomercials, for example. Children see death, assault, and abuse with no feeling of pain or empathy. Yet, TV portrays the real world as a frightening place, full of threats, disasters, and crime.

Despite their power to intoxicate, TV images are illusions. Electrons light up the screen one electron at a time. The electron beam sweeps across the screen and down, covering the whole screen in less than one-thirtieth of a second; all 360 lines are combined by the brain into a coherent whole and interpreted as a single image. Amplified electromagnetic waves picked up by satellite or antenna instruct the TV when to turn the electron beam off and on, creating a pattern we interpret as Lawrence Welk, Howdy Doody, or Dean Martin; but it's really our brains that make the images.

Every phenomenon from the electronic media contains jumps, cracks, leaps, and breaks that contrast with the harmony of the woods that had influenced my first eight years of life.

Dusk had settled in among the oak trees, and the nighthawks darting among the branches chirped like electric sparks. My parents and I were drifting apart. They didn't live up to the parents I had seen mythified on TV—Ozzie and Harriet or Donna Reed, for example. At the same time, my parents were unwilling to acknowledge the youth movement that sought to rip us free of the media. The demonstration in Washington had showed me that the young people of America were out to reclaim their integrity and take back the power of self-identity. It was better to march in the streets, demonstrate, be arrested, feel strong about something, than to stay home and watch TV.

I watched TV for a while that night, and with a mournful sense, I realized I had to wean myself, to leave home for the last time.

Maggi was the only thing I could hold on to. I telephoned her from the desk in the front hallway. I scrutinized the photos hanging on the wall over the desk and listened to the phone ringing in the Bronx. The only photo of me, in my Boy Scout uniform with the Eagle badge, was more than ten years old. There were new photos of my sister's children, my brother's daughter, and a framed photo of my mother and father, from the *Lynn Evening Item*, the day they heard the news we had reached Khumjung.

Maggi was different from the members of my family. She and I met and fell in love in the utter poverty conditions of Peru. There is nothing more real than poverty.

"How's it going?" Maggi was asking.

"Bad. I don't have a home any longer. I'm coming to New York. Can you meet me tomorrow?"

"Ah. Tomorrow? Ah, no, I have a date with Roger Whitehouse," she replied.

"Roger Whitehouse?" I paused, as the impact set in, and then ventured, "Are you sleeping with him?"

"Yes. . . . I'm sorry. I can't stand it," she replied hesitantly. "I'm so confused."

"Oh, shit." I pulled out my bottle of Valium and swallowed one.

"I'm so sorry. Call me when you're in New York. I want to know we are still friends," she said.

"Yeah, oh, sure. I'll call you." I hung up the phone. Damn!

My parents had gone to bed. I turned the TV on and tuned in *The Tonight Show*. I watched until one, and the station signed off. I sat for a long time staring at the white snow dancing across the screen. I knew one out of ten dots was generated by photons from the background radiation, the leftover light from the big bang stretched out to the wavelength of TV waves by the expansion of the universe, downshifted like the moan of the midnight special.

Question: "What's on TV after 'The Star-Spangled Banner' signs off?"

Answer: "The big bang."

Everything in the universe is speeding away from everything else. I had never felt so alone. I stared at the TV as though it were a campfire; a cold glow flickered through the thick darkness and reflected on the metal screens walling in the porch. The pattern of snow on the tube was random, but I could project the sense that the dots flowed toward the center of the screen. I reversed the flow and the dots moved outward from the center. If I moved my eyes fast enough, plays of blue and red erupted across the screen like the northern lights.

I walked the lawn as the night crawlers emerged in the moist grass. The light in my parents' room went out. The dusky cry of a whippoorwill ignited in its hidden nest of brown leaves and sent a shiver up my spine. There is nothing sadder. Who knew if I would ever hear it again where I was going—wherever that was. The dormer windows and roofline loomed dark against the brightness of the Milky Way. The silhouettes of black-oak branches framed Cygnus, the Swan, pointing south toward New York.

The next morning I packed up and looked around my bedroom at the cross-country track trophies and the birds carved by my grandfather.

I yelled down the cellar stairs to my father. "I'm going now, Dad. See you."

Without coming upstairs, he yelled back, "Okay, see you."

In the kitchen, my mother was leaning over a sink full of breakfast dishes. "See you, Mom." I kissed her on her cheek.

She looked up and said, "Okay, Roger, I'll be seeing you." She patted me on the chest and went back to washing the dishes. I didn't know it at the time, but that was the last time I would ever kiss her.

The door chime rang out, "*There's no place like home.*" Yeah, there's no place like nothing left to lose. The flag rippled pleasantly in a stiff breeze.

I stopped at the base of the driveway, gathered myself together, and headed for the bus stop down the street. I passed the suburb's smooth linear lawns and rectilinear walls while playing my bamboo flute, a wild, haunting huayno from Peru. Neighbors pulled lace room curtains aside in the black windows of the houses around me. Doors opened; neighbors and relatives glanced out at me.

Dick McDermott came out of his house. "What's going on? What's all the racket?"

"Oh, nothing," I replied with a slight smile, staring straight ahead.

He closed the door and disappeared inside.

It was May 29, 1969, New York City. The whores, druggies, all the bad wanderers, horny for new illusions, were prowling free on Forty-second Street, seeking shadows of love, jerking off in peep-show fantasies, standing in sticky-soled moats peering up through shuttered windows, coin-operated Automats of lust. Nude women pushed their crotches against the glass windows while speaking to each other of sweethearts and children in New Jersey.

Moral decadence was my new home; the lack of pretense relaxed me. The people of the street were united in their desperation—no hope of their fantasies coming true. They had given up on the squalid American dream—the gross miscalculation that every man, woman, and child on earth, from Bangladesh to Paraguay, could live in suburbia with a Ford in every garage. There was more truth on Forty-second Street than in the lecture halls of Yale. In a society that banishes its citizens to the streets unjustly, the place for the just is the streets. The smashing of my established reality left me with no fantasies to cling to. I was experiencing an emotional death beyond my wildest fears.

I gave up on transcendence. Who knows? Perhaps Thoreau was the ultimate con artist, the P.T. Barnum of the deciduous forest. Perhaps transcendence was rich men's hype, reserved for the affluent, morally stringent New England families who could ransom a piece of paradise lined with stone walls and colorful maple leaves. The working stiffs of the world, the parents struggling to support their families, did not have time or money to seek enlightenment. Is there transcendence on Forty-second Street? I didn't care one way or the other.

Supper in Tad's Steak House did not cheer me up. My despair was deeper than low blood sugar. What meat was being butchered in the night kitchens of America? I took a Valium, and stepped back out onto the sidewalk. I thought I caught sight of Maggi out of the corner of my eye, darting down a Forty-second Street subway entrance with Roger Whitehouse. I imagined my rival was an incredibly cool bohemian with beard and long hair, like a rock star, like Frank Zappa.

Bright letters on the old Times building spelled out the body count in Vietnam, an illusion of marching letters created by switching the bulbs off and on in sequence. Letters formed words that, in turn, formed sentences that marched across the facade. I defocused my eyes; the flashing lights became random, began to move backward.

Chombi claimed all is illusion; like Einstein, I clung to the hope there was a deep reality beneath the surface. But the reality ingrained in me during my childhood was being swept away. I grabbed at every twig and branch as the banks of the river rushed past, finally shoaled, and hauled out among the sins of the street.

I stared, mesmerized, at the illusion of the marching death count. Thank God, I could never kill anyone to protect the reality of a society that had only contempt for me. The feeling's mutual, I'm sure. My father was a hero that I could never be, because there was nothing I believed in enough to kill for. Perhaps Chombi was right: the only way we could be alive was by putting ourselves in death's way. Perhaps everybody is after his or her own near-death experience but has to cloak it in moral justifications and rectitude.

The entire consumer culture is in denial about death, a society of necrophobiacs who believe that death can be bought off, until it overtakes them catastrophically. The old and dying are shunned and shamed. Elders are suspected of suffering from mental incompetence and bad body odor, made the butt of jokes on TV sitcoms.

What good were my near-death experiences? They had convinced me that it was better to be dead than alive, that some part of us, a higher mind or whatever, survived after death. Why was I alive? I was undergoing a kind of slow-motion near-death experience; my normal perceptions of the world were peeling off as surely as if I were careening down the ice cliffs of Mount Everest.

The stairway of the Times Square subway station, walls tiled white like a men's room, stank from stale urine soaked into the concrete floor. The trains roared back and forth on the express tracks like flying gunships strafing rice paddies. I thought I saw Maggi speeding by on a downtown express.

When a local, labeled URANUS II in fluorescent spray paint, finally stopped, the screech of the brakes on the rails, like fingernails on a blackboard, sent a convulsion up my back. I boarded and rocked south to Greenwich Village. The fluorescent lights blinked and coiled back and forth through long tubes as sparks jumped off the third rail. The riders, sick and pale, stared at the floor abstractedly, not making eye contact, because to make eye contact was to open up to the possibility of weirdness. But in New York, staring at the floor was not enough.

A blind man tapping a cane stumbled through the cars rocking back and forth. He took up position on the uptown end of the car and sang, *"If you tear this house down! . . . You can tear this house down!"* The fluorescent lights in the car blinked off, and when they came on again, the blues singer was gone.

As I walked up the steps of the Fourteenth Street station, I studied the faces of the crowd rushing down the stairs and pushing through the turnstiles. I expected to see Maggi in the crowd. My normal thoughts of expecting the worst or feeling depressed bounced around at a shallow, ephemeral level. The thought of bumping into her dawned from the deeper levels of my mind, a growing certain precognition that I was about to bump into her. I didn't think twice about how unlikely it would be to accidentally run into her in a city of seven million people. I just knew that I would.

I wandered aimlessly. I turned south on Fifth Avenue, then east on Eighth Street and passed a movie theater. I turned north on the next block, then west. Sycamore trees shadowed factory lofts. A searing bead of red light traced down my arm, so intense that it seemed to go right through my hand. This was the first laser I had ever seen. I knew they were hypothetically possible, a direct outcome of quantum mechanics. After close contact, all individual waves of light became phase-entangled, or pumped, acting as a single coherent wave. I traced the beam through the fog of steam vents and car exhausts to a second-story window above a cloth banner advertising an art exhibit in an upstairs gallery. I climbed the stairs to the second floor. I'd never seen any art show like it. A large waist-high, glass-enclosed table continuously played billowing shapes of silver and black, like an erupting volcano or towering thunderheads. Currents of thick multicolored liquid flowed together in whirlpools, streamers, and waves, the same dynamic shapes that appear at all scales of nature as dissipative and self-organizing structures; science was just beginning to take note of what the art world had already discovered.

Sculptures of spray-painted wire-mesh screens revealed patterns of changing interference as I walked around them, curving, twisting, and growing apart as the distance between them changed. I had noticed moiré patterns before. For example, when driving past two parallel picket fences, a third pattern emerges. I was seeing an analogy to the basic hidden order of the universe, quantum waves constructively interfering to produce enhanced shapes here or destructively interfering to produce empty space there.

The gallery brightened and I looked up, expecting to see Maggi come through the entrance. Thoughts of meeting her came spontaneously now, without effort. I couldn't stop them if I tried. I wasn't sure why, but they were overpowering.

I was unusually calm. The schemes and plans that normally careened around in my brain had stopped. I had given up hope of anything ever going my way again. I had nothing left to lose, a sense of sureness beyond the chemical effects of Valium. I decided to walk toward the East River. I realized I was headed in the wrong direction, south, toward Battery Park. I turned east and, when I looked up, found myself at the Eighth Street Cinema for the second time.

I bought a ticket to see the movie *If,* starring Malcolm McDowell, and sat on the floor of the lobby with other young people. As the early show let out, I watched the crowd pushing through the doors from the interior to the lobby. The crowd slowed down, and I expected to see Maggi step through the door. She didn't appear. I went inside, and sat in the right-side aisle seat five rows back from the screen. The two seats next to me were empty. As the houselights went down, rows of tiny lights framing the aisles emerged and the faces of the audience fell into darkness. The movie started, and the reflected light from the screen flickered dully across the audience in the first rows. I focused on the movie and I forgot the precognition of meeting Maggi. A boys' choir sang, and a saying flashed on the screen:

Wisdom Is the Principal Thing;
Therefore Get Wisdom:
And With All Thy Getting
Get Understanding.

After the titles ran above the skyline of a British boarding school, the scene opened in the corridors of the school with boys in black uniforms pushing and shoving luggage and furniture. The student proctor, looking very mod, stared at a bulletin board. A lower classman addressed him, "Excuse me. I can't see my name. I'm new."

"You don't speak to us. You're scum, aren't you?" the proctor replied.

A man and a woman in shadows rushed down the theater aisle and stopped beside me. The seats next to me were the only empty seats in the theater. I swung my legs to the side so the woman could slip past. I immediately knew it was Maggi. My first reaction was to look away and deny what was happening, in the hope that by denying it, it would go away or wasn't there in the first place.

"You're blocking my view, scum."

My second thought was to escape before they realized what was happening. If they didn't know it was me, did it really happen? As Roger Whitehouse slid into the seat next to me, I did neither, but reached across him and tapped Maggi on the knee with my flute as she stared up at the screen.

I leaned forward. "Hey, Maggi."

She looked at me, shocked; her mouth opened, and she put her hand over it as if to block a scream. "Oh, my God! Tell me this isn't happening."

"I'm afraid it is. Sorry." I sat back and looked up at the screen.

She looked at the screen for a while, looked at me, swallowed hard, and said, "Oh, well, Roger Whitehouse meet Roger Hart."

"Roger Hart?" He politely shook my hand but withdrew his abruptly and stared up at the screen. We all stared up at the screen. He leaned toward her and whispered something. They both stood up. I swung my legs to the side without taking my eyes off the screen. They squeezed out in front of me without a word, and soon were gone as if they had never been there.

On the screen, the boys unpacked their belongings in their dormitory. Malcolm McDowell hacked at his newly grown mustache.

"God, you're ugly. You look evil," said a classmate.

"My face is a never fading source of wonder," said Malcolm McDowell.

"Why did you grow it?"

"To hide my sins."

My normal thought litany returned, and with each moment that passed, I became more convinced that what had just happened had not happened. It had almost disappeared from existence when Maggi rematerialized, kneeling beside me in the aisle.

"Roger, we need to talk. Can we go someplace and talk, the three of us?"

I felt an immense sense of calm and power as I followed Maggi back through the audience in the dark theater.

"Fantastic!" was the last line I heard from the movie *If.*

Roger Whitehouse waited in the lobby, looking like a medieval gnome, his hair and beard long and scraggy. He spoke politely with an English accent. "Would it be okay if we talked?"

"Yeah, sure," I said.

"It's quiet in the upstairs lobby," said Maggi.

We climbed the curved, carpeted stairs to the mezzanine lobby. The doors to the balcony were closed, and we were alone. I sat down on a vinyl-covered bench. Maggi leaned against the wall.

Roger sat down beside me, lit up a cigarette, and looked at me, knots of concern in his forehead. "You know how difficult this is on Maggi?"

"Yes, of course it's hard on Maggi; I can't help it, I love her."

"For her sake, she's got to decide between us."

"I know it's tough on her, but I'm not forcing her to decide between us," I replied.

"This is ridiculous. The two of you broke up. And now I can't go out with her without running into you."

"It was an accident," I said.

"An accident? You mean the two of you didn't set this up?" he asked.

"No, we didn't set it up," said Maggi, a little flustered. "But I don't think it was an accident either."

"Perhaps not," I responded. "Seven million people in New York, hundreds of movie theaters, what's the chance of an accidental meeting like this?"

"It's a miracle," said Maggi. "Look, I gotta go." And she ran off down the stairs.

I had built up a picture of Roger Whitehouse as someone much cooler and more sophisticated than me. But to actually run into him popped my distortions.

"I'm just concerned for Maggi," he said. "There's the question of her parents; she can't leave her parents and go running off to Latin America with you."

"I know you're right," I said. "I just know I love her. That's why I'm here. I didn't know she was seeing you when I came back. I don't want to make demands on her."

There was a long silence.

"We'll just see what happens." He stood up slowly without looking at me and started down the stairs.

"Yeah, we'll just see what happens," I replied.

As soon as he disappeared into the crowd on the first floor, I was ecstatic. Not because I had vanquished a rival lover—I wasn't sure I had done that—but because it was clear my life was in the hands of a higher power. To have been part of a miracle implied a greater purpose to life. More important than this incredible coincidence, this miracle, was that I knew in advance it was going to happen.

Perhaps, though, I was just remembering that I foresaw it, and my memories were incorrect. The really important thing was that, whether or not I foresaw it, the meeting had happened and I felt great. I was sure of that. Yet, the more I thought about what had happened, the more the wonderful feeling went away. I decided to not think about it anymore.

I stared up at the tops of the airy sidewalk sycamores, exuberant as I worked my way back to the Fourteenth Street station, light in my footsteps like after dropping a heavy backpack. All my worries about life, family, livelihood, had vaporized. A glowing neon light proclaimed "Electric Circus" on St. Marks Place. I popped in for a minute. Strobe lights flashed across the dance floor; Janis Joplin and Big Brother and the Holding Company were on the turntable singing "Piece of My Heart." The strobes stimulated my synapses, filled me with energy. I couldn't have stood still if I wanted to. I spun, hopped, and danced. A whole new life had opened up as my old one fell away. There was a long-term plan, a jubilant goal I didn't understand. I didn't care what happened to me; it was just good to be alive. Everything was perfect the way it was.

After several hours of dancing, I strolled to the Fourteenth Street subway station. The subway glided into the station; I rode it uptown like an amusement park ride, smiled at everybody.

The lights of the building at One Times Square no longer formed letters and words. The message was irrelevant; the number of dead in Southeast Asia didn't matter. To believe it, to react to it, was to give up power. The flashing lights were the message, a worldview based on electromagnetic radiation. I stared at the random patterns of lights until the space around me scintillated with light dots at a speed that sent shivers up my spine. With perfect clarity I understood I was projecting meaning onto the fabric of reality popping out of some unknown background energy, one flash of light at a time. All former perceptions of reality seemed irrelevant.

The experience of running into Maggi in the movie theater was different from my experiences on Mount Everest and in Tierra del Fuego. I *knew* that it was not a coincidence, and it had the added element of precognition. How was it possible to see the future within the normal time framework of everyday life? That was the question. Somehow the future must be accessible from the present. The usual answer I had heard is that time is an illusion. But what makes it an illusion, why does it seem so real to us, and is there some level of existence that is free of the illusion? I was sure that if I could answer those questions, I would discover something fundamental about the nature of reality. Even if I could demonstrate that there is a place where time does not exist, there was a more fundamental question—that of free will. The ability to see the future implies lack of free will, a future cast in stone, a predetermined (boring) universe. I had no idea how to solve this problem.

Even though the emotional death in New York was not as catastrophic as the near-death experiences on Mount Everest and in the blizzard in Tierra del Fuego, the internal conditions were the same; my normal mind was emptied, providing access to mental states that were focused and apparently connected to other objects and beings. But how was that possible? I didn't have a clue and tried to forget the experiences.

Despite my doubts, the experiences were fascinating, taught me something directly, and I wouldn't have them any other way. I trusted that they had happened for a reason that would eventually become clear. Otherwise, I should just forget them; however, each time I tried to forget them, a new experience arose, bringing more complexity to the mystery.

Later that same summer, I attended an extraordinary event that confirmed many aspects of the near-death experience I had been part of in Tierra del Fuego.

I met Maggi in Central Park. "It's supposed to be the greatest rock-and-roll show ever," said Maggi. "Let's go."

"Ah—Is Roger Whitehouse going?" I asked.

"No, he's on a road trip, probably in New Orleans by now."

We took a Greyhound from the Port Authority bus station to the Catskills and spent the night in a hotel in Monticello. The next day we hitchhiked to the festival grounds. Traffic was backed up, row after row of cars were parked helter-skelter along the road and in the fields. A car leaving slowed down, and a girl in the passenger seat rolled down her window, and yelled to us, "Are you guys going in?"

"Yeah, we're trying," I said.

"Take this." She handed us a car-window PRESS pass. "Pass it on."

I held up the pass over my head, and the very next car going in, a '55 Chevy, stopped. Thunder peeled out and rain poured down in sheets. The traffic controllers in white armbands stenciled with a dove directed us around a roadblock into a service road. The car bounced through muddy ruts; the wipers slapped at the windshield. Thousands had already left the Woodstock festival, a river of teenage youth straight out of middle-class suburbia, sopping wet and covered with mud. Like dog soldiers in Vietnam, they slogged through the mud lugging knapsacks and sleeping bags, oblivious of appearance. Laughing and singing, they marched with stringy and matted hair. Clothes stuck to their limbs as if they were in a wet T-shirt contest that was evolving into a no-T-shirt butt-naked contest, holding sheets of white plastic over their heads. The children of the suburbs were enduring, albeit temporarily, the hardships of the poorest Third World culture. Instead of watching the incessant replays and analyses of the first moon landing and Kopechne drowning on TV, these children of America chose the mud and discomforts of Woodstock. In fact they seemed oblivious to the discomforts. We pulled into the middle of the concert area and hopped out of the car. Some youths were running up to a long runway of mud and launching themselves onto it on sheets of cardboard and sliding into each other.

The chain-link fence had long been trampled into the ground, the box office closed, and the grass field transformed to mud. Joe Cocker had just finished singing "With a Little Help from My Friends."

"It looks like we're going to get a little bit of rain, so cover up. Hold on to your neighbor. You gotta get hold of your friends now," Wavy Gravy yelled into the speaker system.

The scene was dismal—drab, gray, no color, only white T-shirts and jeans like a crowd at a high school football game. People were wrapped in blankets, some shared blankets, and others huddled under sheets of white plastic that also covered the muddy field.

One of them saw me staring.

"Hey, brother," he said.

Let's get out of here, I thought.

"Here, brother, want some bread?" He held out a loaf of French bread.

"Yeah, sure, thanks." I shared the bread and joint with Maggi. I felt out of place. What was I supposed to do? *What's going on, am I missing something, why is everybody else having a better time than me?*

Country Joe and the Fish set up onstage. The cranes and sound towers hovered over the muddy meadow like a construction site or a missile launchpad, minarets reaching skyward from some other world. The noise of the pounding electric guitars appalled me. This was nothing like Lawrence Welk or Dean Martin. Charging rhythms and screeching chords wailed and twisted on the tight wire of discordance. This was a new use for the electromagnetic spectrum.

The wind whipped up the tarps covering the main stage. Nimbus clouds rolled in over the fields

"Think hard to get rid of it," yelled Wavy Gravy.

"No rain! No rain!" The crowd picked up the chant.

The stagehands scurried on the main stage unplugging and covering the equipment. The pot kicked in.

I heard the rhythm of bottles, cans, and sticks, which reminded me of the samba schools in Rio during Carnival. Maggi and I pushed through throngs dancing wildly, beating bottles, sticks, and cans: *Dat dat dat-dat dat dat-dat, dat, dat. Yah! yooh oho oh o.* It was pouring down rain, but everybody was smiling, hopping, dancing, and spinning. Bongos rattled in the background like jungle drums. The writhing dancers beat out a rain-forest tribal rhythm, chanting: *Hey heyey hey! Hey heyey hey! Boom chak-a-lak-aah.*

The rhythm captured me and built to a crescendo. I spun around Maggi, and she spun around me, smiling, laughing, and looking beautiful. I snapped pictures and bobbed and wove in step with the dancers. By taking pictures, I was probably missing out on something great. *Peace peace peace peace peace peace peace.* We chanted like a locomotive working up to speed.

In my wildest thoughts I did not expect this, could not have even imagined it—in upstate New York! It was not the worst or the best but beyond judgment. My mind stopped thinking, free of words. Wild drumbeats invoked the raw power of nature, pagan hearts ensconced in primal and forbidden forces, free of doubt, fear, and guilt.

I felt someone hug me and drape an arm over my shoulder. It was Jeremy, looking handsome, his own musketeer-self in a red headband.

"Jeremy, I haven't seen you since the D.C. jail. What a coincidence!" I yelled over the din of drums. We danced together arms over our shoulders.

"A coincidence? No way." He smiled mysteriously and handed me a joint. I took a drag. He spun away.

The rhythm entrained our minds into a single wave of energy, floating in a golden haze a few feet above the ground and separated from our bodies, one mind in coherence, joyful and beautiful, beyond any recognition of time or place.

The drabness lifted, the scene filled with light, and sunlight exploded from below the clouds and gilded the muddy field. The dancers threw their arms into the air spontaneously, jubilant.

Jeremy yelled out, "Look! We stopped the rain!" He threw up his arms to the sun.

We did? I thought to myself. With that question, I found myself prancing ridiculously in the mud, alone, cut off.

The other dancers spun with their arms in the air and I took their picture. *It was just a coincidence.*

At a watershed of cultures, the ultrasonic rhythms dismantled the old generations and revealed the mysteries of complexity. I felt accepted, alive, rejuvenated, ready to set out on a new life rising like a phoenix out of the ashes of arrogance, a new world with no presuppositions of right or wrong, no judgment or blame—a world, I was sure, that would include me. The icons of the older generation were being blown sky high, and in that moment of release, miracles were happening. The pieces were settling into a new reality.

By the end of the summer of '69, it became clear I was stuck in the States and needed to find a job. I accepted a position with the Naval Oceanographic Office in Suitland, Maryland, and joined project GOFAR on an outpost at Chesapeake Beach. For the first six months, Maggi flew from New York to visit me on weekends. In early 1970 she accepted a position as director of a Head Start center and we moved into an old farmhouse on one of the tributaries of the Potomac River.

I needed a focus for my research, so I talked to my brother, Stanley, who at the time was at the Department of Terrestrial Magnetism of the Carnegie Institute, in Washington, D.C. The theory of plate tectonics had just taken off, and Stan's thesis adviser at MIT had helped develop the theory. Certain elements have more than one form, known as isotopes, varying subtly in weight but not in chemical properties. Mass spectrometers were used to measure the composition of isotopes of strontium in rocks from Africa and South America. A clear lineation of isotope trends in Africa could be matched up with one in South America, confirming that the two continents were once joined together like pieces of a jigsaw puzzle. The continents had moved apart, at the same rate at which a fingernail grows, on slabs of crustal rock that separated in the middle of the Atlantic Ocean basin along a rift system populated with volcanoes. Similarly, the Indian subcontinent has been moving north into Tibet and pushing up the Himalayan Mountains.

Stan was using a mass spectrometer to study the chemical composition of volcanic rocks dredged from the Mid-Atlantic Ridge. He found that the outside of rock fragments had exchanged chemicals with seawater. He proposed that I carry out a statistical analysis of the published chemical compositions of dredged volcanic rocks. I enlisted the help of computer programmers at project GOFAR. We plotted parameters of chemical and physical properties in large printouts that I plastered over the walls of my office. I studied the plots as eagerly as if they had been maps of uncharted lands. I submitted the data to my unconsciousness, which over time ascertained relationships that seemed to jump into consciousness unbidden. I found that rocks varied in their chemical composition with the distance from their volcanoes of origin. The chemical trends were the same as those Stan had found comparing the inside and the outside of single samples. The conclusion was that the rocks aged as they moved away from the spreading ridge, and as they did so, they exchanged chemicals with seawater.

Increasingly in science there is recognition of the *eureka moment*. This is the moment when a new discovery, a level of understanding, or a major revelation comes into the light of consciousness for the first time and never fades away. It is personally exhilarating, almost addictive. Once it happens, the investigator exhaustively seeks, sometimes futilely, after new occurrences as if trying to score illicit drugs on the street. The *eureka moment* came for me when I calculated that there was enough chemical exchange going on for this process to be influencing the chemistry of seawater. It came as an intuitive flash when I least expected it. As far as I know, I was the first to have this realization, and it was very exciting for me. I published the results, with the title

"Chemical Exchange Between Sea Water and Deep Ocean Basalts," in *Earth and Planetary Science Letters* in 1970.

Several years later I made a similar study of rocks that had interacted with seawater at high temperatures near magma chambers deep in the oceanic crust and published the results in the journal *Nature*. These papers were well received and were my first experience of the satisfaction of intellectual discovery.

Eventually deep-diving submersibles found reaction centers, hot springs, where heated seawater laden with minerals exited so-called black smokers. Strange new life forms living on the energy of chemical reactions, rather than sunlight, were discovered. I was amazed and pleased to note how a simple study of volcanic rocks had blossomed into a major new area of research. The experience was gratifying and taught me that the slightest new observation can lead to exciting new discoveries.

Despite the success of my scientific work, my tenure at the Naval Oceanographic Office was short-lived. I was scheduled for a cruise off the Galapagos Islands that required Department of Defense clearance and a DOD passport. When it was discovered that I had been arrested during a peace demonstration, my DOD clearance and passport were denied. My superiors at the Naval Oceanographic Office did kindly expedite a new civilian passport from the State Department. On the basis of my published papers, I was offered a temporary position at Oregon State University and, eventually, at the Physical Research Laboratory in Ahmedabad, India.

In the meantime, Maggi and I decided to take a trip to Morocco. It was there that the most profound experience of all occurred.

CHAPTER SEVEN

MAGIC COOKIES

A yellow moon hovered on the horizon as our bus rattled up to the sandstone gates of Marrakech. The silhouettes of palm trees and minarets reached into the purple sky like the fingers of Allah drawing down the curtain of night. The bus wallowed through throngs of hooded figures and lurched to a halt at the main plaza. As Maggi and I climbed down, I clutched my shoulder bag containing our money and passports. The screech of a snake charmer's *shehnai* careened off the walls of the mosque. The crowd applauded and cheered as acrobats spun, twirled, and flipped in white briefs and red-silk waistbands.

"You want hash cookie, very nice?" A man in a brown-hooded djellaba stepped out of the shadows into the light of the hissing white gas lanterns.

I scanned his dark face. "What do you think, our first rip-off?" I whispered to Maggi.

"Hmmm, he has kind eyes." Then she shouted above the din of the plaza, "Did you make the cookies yourself?"

"No!" He laughed, rivers of wrinkles cascading over his cheeks. "My daughter make them; I sell them."

"How old is your daughter?" She was warming up to the man; she liked anybody with children.

The man held up ten fingers as the light of the lanterns sparkled on his silvery beard stubble.

"Ten? Amazing!" Her childlike eyes widened with curiosity. "Where do you live?"

The man pointed to the shadowed archway that led to the medina. The medinas of North Africa, the Casbahs in the center of the cities, are intricate labyrinths of narrow streets. I had heard stories of Westerners robbed in the medina who are still trying to find their way out. I leaned toward her and said firmly, "Maggi, we are not going to the medina with him."

"I think he's okay; don't worry." She pushed me back with her hand.

"How much are the cookies?" She smiled and stepped close to the man.

"Four dirham. Come! Come!" He glanced about, placed one hand inside his djellaba, and motioned with the other into the shadows behind the food stalls. A burst of flame leapt up from a row of shish kebabs, and the smell of charcoaled lamb curled into the shadows. He pulled out a paper bag and held it under Maggi's nose.

"Umm, smells nice!" She fished her embroidered change purse out of her shoulder bag and counted out four dirham into his gnarled hand.

"You come! My house tomorrow! My daughter show you how to make very nice hash cookie." He pointed to the medina. Then he disappeared into the crowd.

Maggi stuffed the paper sack of cookies in between her rolled-up orange windbreaker and rumpled blue jeans. It always amazed me how much she could cram into that string bag. It bounced off her hips as we wove our way across the plaza. A beggar grabbed the hem of my black cape. I pushed him off. Hawkers screamed and waved their wares at us: avocados from Agadir, dates from Sidi Ifri, fragrant mint from Tétuoan. Bumped, shoved, and feeling anxious, I grabbed Maggi's hand, and quickened our steps. We desperately needed to get out of the crowd, find a hotel, and relax. A blind man, hunched on the pavement, twanged at a stringed instrument and wailed into the night like the Sahara wind. A woman beside him swayed on her haunches, veil arched up beneath kohl-lined eyes, and clinked out the rhythms of the desert on finger cymbals.

"Baksheesh! Baksheesh! Baksheesh, memsahib," she screamed, and pointed to the blind man's turban on the sidewalk.

Not tonight, I thought to myself, and pushed on by.

"Look, there!" Maggi pointed to a neon sign that flashed

FREAK OUT SNACK BAR! "Where there's freaks, there's hope. Must be a hotel nearby."

Sure enough, a small sign jutted out into the street. "EUROPEAN ACCOMMODATIONS," it read.

"What do you suppose 'European Accommodations' means?" I pressed the night bell.

Maggi shrugged. "Beds, maybe."

"Beds with ropes or beds with springs? that is the question."

A matronly woman cracked opened the door and peered out over her gray veil. She gasped in shock at the tangle of my black hair and pushed the door shut as if I were the devil himself.

"I hope it's springs." Maggi stepped in front of me. *"S'il vous plaît, madame, une chambre pour deux jours?"* She pushed gently on the door.

The door opened a crack, and the woman eyed Maggi up and down. *"Une chambre avec un lit?"*

"Oui, oui, un lit."

As the woman dragged the heavy wooden door open, a black man in a flowing white cape and cowboy boots rushed out of the Freak Out. Looking as if he was straight from the East Village, he wore a cape like mine, only white. He veered toward me, offering a thumbs-up handshake. "Hey, brother, where'd you get your cape at?"

"Medina at Fez." I shook his hand.

"No shit, man, me too." He looked me up and down. "Hey, man, didn't I see you at the Fillmore Dead concert?"

"Nope."

" 'lectric Circus?"

"Possibly."

"Aw, shit, man, where you coming from?"

"Boston," I replied defensively.

"Oh, no, don't give me that shit. Boston? You got to be that uptight." He waved his hand at me in disgust, turned, and trotted off down the street.

The exchange apparently confirmed our landlady's worst suspicions; she scowled at me as she led us through an open patio and up worn wooden stairs to the third story. Through the open skylight, I saw that the purple had left the sky and the moon glowed coldly, like silver against crushed velvet.

The landlady opened a glass-paneled door with a brass key, flicked on the light, and swept her arm across the room.

"Tout bon?" She smiled at Maggi and turned to leave.

"Tout bon." Maggi smiled back at the landlady as she closed the door behind her. The room had a wooden floor and smelled of hemp. The paint was peeling off the walls. Cracks and water stains mosaicked the plaster ceiling.

I dropped my bag on a table at the head of the bed and lifted up a corner of the mattress. "Ropes!"

"Oh, well, a few more nights on hemp won't hurt us." Maggi sighed.

I spread my black cape out over the bed, lit a candle, dripped some wax on the table, and stood the candle upright in it. Maggi lit sandalwood incense and wafted the smoke around the room like a choirboy at mass, the smell overpowering that of the hemp.

"That'll get rid of the bad vibes." Maggi turned off the overhead light. The warm light of the candle dawned into the room, and the cracked walls disappeared into shadow.

We embraced and kissed in the middle of the room. "Feeling better now?" she asked.

"Yeah! Home away from home." We had performed the same rituals in cheap hotels from Dublin to Granada. "Want to smoke?" I pulled a kef pouch from my bag.

"Yeah, sure." When she sat down on the bed, the wood frame creaked. She pulled off her pink embroidered blouse, brown corduroy pants, then her panties. Her flesh glowed with an orange light. I filled the pipe, undressed, and sat down beside her.

"Well, no rip-offs yet," I said, sucking deeply on the pipe. The flame flipped to the bottom of the match. I held my breath and watched the flame flip back to the top of the match as I passed the pipe to her. I let the smoke hiss out. "Can't believe the stories we hear on the road, nothing but paranoia."

"I know; people pick up on the bad vibes, only makes things worse." As she sucked on the pipe, she arched her back. The freckles on her arms and shoulders framed her soft breasts, smooth as cream in the candlelight. She exhaled and set the pipe back on the table. I brushed the silk of her breasts with my lips. A shudder went through her body. She wrapped both arms around my head and pulled me tight. As we sprawled backward onto the bed, the ropes groaned like a windjammer tethered at the turn of tide. A moan vibrated on my lips against the flesh of her neck.

After making love, we lay on our backs in bed, arms under each other's heads, my leg between hers. The candle threw a steady orange glow across the cracks and circular rain stains of the ceiling. Maggi stared at the patterns, her face still flushed with the lovemaking. She turned to me with a soft smile.

"You ever drop acid?"

"No, I'm afraid to; you hear so many scare stories about bad acid. But I took yage with a Shipibo shaman in the Amazon. You know, the *Yage Letters*, Burroughs and Ginsberg?"

"Yeah, acid and yage," she replied. "Both are hallucinogens."

"Hallucinogens, I'll say. I took yage in this ceremony with some village Shipibos. I was supposed to meet my spirit ally—you know, a jaguar or panther, a totem, something that would help me with the trials of life."

"And?" she asked.

"Do you remember flatworms from high school biology?" I asked. "You know, *platyhelminthes*? Planaria?"

"Sure the cute little cross-eyed worms." She crossed her eyes and undulated like a flatworm.

"Well, after barfing and rolling around on the ground for half an hour, I felt as if I was split into four or five different levels. I was freaking out because I didn't know which level was the real me and I didn't know if I'd ever get back to the right level."

"Yeah, that happens on acid too," she replied. "And don't worry about it; there's no right level."

"Just when I was feeling the worst, I looked up and saw this giant creature arching to me out of the sky, a huge flatworm. I mean it was real and scared me, but at the same time it was all in my head. And that freaked me out me even more, to think of things like that in my head, that I might never get it out of my head. I sprawled facedown on the earth like I was half embedded in it. The fear flowed out of me as if I was grounded."

"Planaria, aren't they the ones who grow back after they are cut in half?" asked Maggi.

"Yeah, I think so, even quarters."

"Your spirit ally? Your totem is a flatworm?" She looked at me with an expression of gleeful disbelief.

"I suppose, yeah, sure, why not? My spirit ally is a flatworm."

"Hey that's great. Flatworm guts jaguar. If a jaguar chews a flatworm to pieces, each piece grows into new flatworms, great ally.

How do all those little pieces know how to fill out the new flat-worm shape?" She shook her head and laughed.

"I don't know," I replied. "It seems like magic."

"Do you believe in magic?"

"I don't know," I replied again. "You must believe in miracles, you're Catholic. Isn't it required?"

"Magic is different from miracles," she said thoughtfully. "I'm talking about magic done with the mind, things influenced by thought. Miracles are done by divine intervention." She had renounced the Catholic Church as materialistic and dogmatic, but she still defended the saints of her childhood.

"I don't see the difference." I picked up my cape from the floor where it had fallen during the lovemaking, rolled it up, and stuck it under my head as a pillow.

"I mean magic done deliberately by a person," she explained. "Like in this Sufi story I'm reading. A monk meditated in a forest. After five months of concentration, he killed a sparrow just by wishing it dead." She pushed her string bag under her head as a pillow.

"Then what happened?" I yawned, slid my arm out from under her neck, crossed my arms on my chest, and shut my eyes.

"I don't know, haven't finished the story yet." She shrugged. "But do you think it's possible to influence things with your mind?"

The midnight prayer call rang out from the minaret at the plaza. I was too tired to respond but wanted to tell her that sometimes it seemed as if the mind could control things, just not all the time. The question was *how* did the mind influence things? Furthermore, I didn't know if it was morally right. That kind of power seemed to violate a basic code of being human. Salem, just north of my hometown Lynn, was famous for prosecuting practitioners of magic. Only Christ and a few saints were supposed to have that kind of power.

Chombi had told me that the wind and weather were the easiest to control—people, practically impossible. I smiled remembering when he had told me to call the wind. I had thought he was joking, but to humor him I raised my arms melodramatically and yelled out, "Come, Wind!" I turned around to see the rhododendrons rustle, and a strange energy surged up my spine. Then I remembered the rhythm of the drums, bottles, and rocks in the rain dance at Woodstock.

I turned sleepily to Maggi. "Perhaps, you've seen magic yourself."

"Really? When?" She raised her head to look at me.

"Remember the rain dance at Woodstock?"

"Yeah, sure." She propped up on her elbows.

"The sun came out during the dance."

"That was just a coincidence."

"That's not what the dancers believed."

"True, they went berserk, but they were stoned, and I don't really believe I had anything to do with that."

I was drifting off to sleep. "Yeah, neither do I." I yawned. "But perhaps we stopped the blizzard in Tierra del Fuego."

"You thought you were going to die." She rolled onto her side facing me.

"That's when magic starts."

"Why then?"

"I don't know."

My head flopped to the side; I lost consciousness for a few moments, then woke up.

Maggi leaned over me, rubbed her breasts against my chest. "Don't you want to make love again?"

"I think I'm too tired; I mean, men aren't like women—we need time to recover." I thought for a minute, then realized I had made a mistake.

It was too late; Maggi was already staring at the ceiling on the other side of the bed. "You only make love when you want to! I need more feeling from you. You might as well be making love to a prostitute or a rubber doll."

"Sorry, I was brought up to believe sex is shameful." I couldn't keep my eyes open. "I mean I've been bouncing around on that bus from Fez all day. I mean, Jesus, what was in that kef?" I lifted my head to look at her. She was facing the wall, a blue funk settled into the hunch of her shoulders.

Didn't satisfy her, the accusation swirled into spirals of blue light in my brain, and I fell asleep.

A hard punch of Maggi's fist against my temple awoke me. Daylight filled the room. "Asshole!" she screamed at me, stomped out of the room, and slammed the door behind. The dawn prayer call echoed from the minarets.

I lifted myself on my elbow and saw her bag on the floor. Not going too far, I thought, and collapsed into sleep. Next, the noon-day prayer call woke me.

Bright light filled the room, and Maggi's bag was gone. Oh, oh, I panicked. This time she'd really left me. I hurriedly dressed, took a leak in the third-floor toilet, ran down the stairs and out into the bright light filling the cobbled street.

I headed for the plaza, but as I passed the Freak Out I saw Maggi sitting in the corner. The black guy was leaning over the table talking earnestly to her. Maggi glanced up as I entered, then bent over her glass of banana au lait. With grave concentration, she sucked up the foam from the sides of the glass.

"Oh, hi there, Boston." East Village gave me a thumbs-up handshake, and motioned for me to sit down. "Where you been at, man? What've you been stoned on?"

"Kef, man." I slid into the wooden booth beside Maggi. She aggressively slurped at the bottom of the glass.

"Kef, you gotta be kidding, kef!" He smiled, flashed his teeth, and slapped his knee. "Wait until you ingest hash; I'm not talking about inhaling a few whiffs. Now, there's an eight-hour trip! Whatssa matter, afraid? Boston, nothing but puritans and blue bloods, right?"

"I haven't been in Boston for a year." I gave him a hard look.

"Oh, yeah? What've you been doing?" He softened as if he was worried that he had pushed me too hard.

"Travelin'."

"Whereabouts?"

"South America—Chile, Ecuador, Peru, that's where Maggi and I met." The sound of the last slurp from the glass was followed by the slip of air.

"Yeah? How you get your bread together?" he said with the feigned concern of politicians and drug dealers.

"She was in the Peace Corps in Peru. I'm a geophysicist, but recently I've been climbing mountains and taking pictures."

"Sounds like a good scam. I stuff hash in bronze statues myself, have a welder friend in the medina. Hey, look, that reminds me, I gotta go."

He stood up, smiled at Maggi. "Be seeing you around the medina, sister. If you change your mind, let me know."

Then he turned to me. "Keep up the old blue blood." And he saluted me as he walked out.

Maggi shoved her glass against the wall and pointed the straw straight out the door after him, a hard expression on her face.

I took a deep breath and looked around the snack bar. Rich maroon caravan weavings hung behind a bright chrome espresso machine. Then I noticed the book of Sufi stories on her lap. "Finish the story about the monk who killed the sparrow with his mind yet?"

No answer; she kept staring.

"Yeah, when you think you're going to die, that's when magic happens," I said.

She looked down at the book in her lap and lifted it onto the table. She tilted the straw down. A drop of banana au lait fell on the white Formica. She drew a yin-yang symbol with the straw.

"I mean, it's like Saint Patrick's terminal fast, when he fought with the snakes."

The corners of her mouth softened. We had climbed to the top of a mountain in Ireland and sat on the spot where Saint Patrick fasted, faced his demons, and assaulted the snakes of the collective unconsciousness.

She ran her thumb over the pages of the Sufi stories and swallowed hard. "The monk ran to town," she said slowly. "He knocked on the door of the first house. An old woman answered. The monk demanded she serve him tea. She refused. He got arrogant and told her, 'Listen, you old hag, do you know who I am?' She replied that indeed she did and that anyone who killed a sparrow didn't deserve a drop of water. The monk realized the old woman knew all that happened and was more powerful than he would ever be. So he spent twelve years as a street sweeper to make up for his pride—and I'm sorry I hit you." She took a deep breath, plopped the straw into the glass, and turned to me, one eyebrow cocked full of doubt.

"I deserved it. I'm a rotten lover," I said, sweeping my sandal across the floor. "I can't get past my inhibitions."

"No, you're not a terrible lover. I just got frustrated and angry." She slid her hand on top of mine.

"I mean, I'm weird." I slid my hand out from under hers. "Whenever I get stoned, I have weird fantasies, all of which are banned in Boston. I've slept with men and women. I'm conflicted and probably always will be."

"You're not gay." She took my hand and held it firm with two hands.

"The Army thought I was, enough to give me a 4-F."

"And you believe them?" she replied. "Nobody believes the Army."

"Well—" I started.

"Bi maybe. Perhaps you're bisexual," she interrupted. "So what? You're like everybody. Everybody has fantasies. We just don't act on them. On the other hand, suppressing them energizes them. Carl Jung got that right. That's what the revolution is about, facing our hang-ups and dealing with them. Talking about our fantasies takes their power away, keeps them from turning into monsters."

"I thought the revolution was about sharing wealth and power with minorities," I said.

"That too, but we have no control over that. Let's start with ourselves."

I looked at her bright, clear eyes, reached up, and brushed her hair back over her forehead. "You're the best lover I ever had," I said to her, and kissed her neck below the earlobe.

She reached under the table, rummaged through her bag, and pulled out the paper sack of hash cookies. "Ready for lunch?" She smiled seductively.

The thought of eating a hash cookie made me nervous. On-the-road stories abound of undercover agents selling drugs to hippies only to turn around and arrest them as soon as they got stoned. Two nights in the D.C. jail had about freaked me out. The jails in Morocco had to be worse. I couldn't handle a Moroccan jail.

"I think I'll duck in the john." I grabbed two cookies in my fist, hid them in my cape, and went into the men's room. My first thought was to flush them down the toilet. Damn, I couldn't handle a bad trip in the middle of Marrakech. I had heard about a girl who got a bad batch of cookies. She went crazy, ripped her clothes off, and started masturbating on the sidewalk of the main plaza. The Moroccan men stood around and watched. Yage dreams in the rain forest were one thing. But Marrekech—what if I was attacked by a worm in the middle of the plaza? I would be the laughingstock of the medina.

And what if Maggi had a bad trip? She never had a bad trip. It was always she that pulled me through. A piece of graffiti above the squat-down toilet read, "Reality is a crutch for those who can't handle drugs." I was an explorer, goddamn it. The only territory left to explore was the human mind. Sure, I could handle drugs. I popped one cookie into my mouth and flushed the other down the toilet.

I immediately felt queasy and I thought maybe I'd made a mistake. Making love would relax me. In my mind, the formula was simple: if I make love, it'll be a good trip; if I don't, it'll be a bad trip. I came out, sat back down at the table. Maggi stuffed the cookie sack back into her net bag under the table.

A Moroccan teenager with close-shorn hair, sandals flopping against his heels, shuffled into the restaurant: either he had lice or he was a "Hare Krishna" looking for baksheesh. He sat on a stool in the front corner, crossed his bare legs, letting his sandals flop loosely, and smiled at Maggi.

Maggi smiled back, and the boy gestured with his hands around his mouth and moaned, hissed, and grunted.

"He's dumb," said Maggi, full of sympathy.

"Yeah, must be if he's a Hare."

"No, no, he can't talk." She brightly signaled back with her fingers.

I turned and smiled at him. He pointed out the door. "What's he saying?" I asked, turning to Maggi.

"He wants to show us the medina. Oh, let's go with him." Maggi perked up and reached under the table for her bag.

"I'm feeling a little queasy. How about we go back to the hotel and rest for a while?" Perhaps we should have sex first, I thought to myself.

"Com'on, where's your spirit of adventure?" Maggi hesitated, thought for a moment. Then her face lit up. "I know. Let's go to that park we saw coming in on the bus." A red blush from the hash cookies had dawned on her cheeks and the tips of her ears. She made little circling motions with her hands toward the mute boy. He nodded his head up and down enthusiastically, constricted his throat, and let out a long grunt.

"What's happening?" I asked.

"I'm telling him that we'll catch him later."

We stepped out onto the street. Dust, noise, and commotion filled the air above the plaza. The sun tilted low and gilded the air with a golden haze.

A man in a lopsided white turban squatted on the ground in

front of an ornate snake basket. He blew into his reed with a loud screech, but no snake came out. The charmer reached into the basket, pulled a cobra's head up six inches, played the reed with one hand, and tapped the snake upward under the chin with the other. At about eighteen inches, the snake opened its pink mouth, spun in a slow circle hissing at the crowd, and then dropped back into the basket. The charmer bounded up, bowing deeply; the crowd roared with laughter and pelted the basket with dirham.

Maggi pointed across the road to a field of palm trees sprinkled round about with a few orange trees. Hell of a park, I thought. No grass, probably full of cobras, leeches, or something. My stomach turned on a pivot of nausea. I wished we had gone to the hotel. We found a patch of skimpy grass between some mounds of dirt. I checked the ground for snake holes, and we lay down side by side with our heads on Maggi's bag. I stared up at the blue sky. The soft sway of the palm branches mellowed me out.

An inchworm spun down a long silk thread that undulated in the soft breeze and splashed off rainbow colors.

Inchworm, a misnomer, I thought to myself, should be 2.54 centimeters. That didn't sound right. Perhaps worms haven't gone metric yet.

The apple-green worm stopped a foot (30.5 centimeters) from Maggi's face, twisted, turned, and arched into different shapes.

Maggi's whole face was energetic pink as she studied the worm, openmouthed. "It's writing out the Arabic alphabet."

"What's it spelling?" I decided to humor her.

"It's spelling out—wait, okay, yeah, it's 'Coca-Cola' in Arabic," she replied earnestly.

"Wow! This worm's looking out for us." I scrutinized it carefully.

"Now he's spelling out something else. Says the mute boy is coming."

Right, I thought, he probably followed us in hope of ripping us off in the medina. We heard a cacophony of moans like a coven of seals barking underwater, and raised our heads to see the mute gesturing in front of an orange tree.

"What's he saying?" I asked Maggi.

"He wants to know if we are ready to see the medina." Maggi propped up on her elbow, turned, and giggled at me with her hand over her mouth.

"I'm not sure, but I'll go if you want to." I decided to humor her some more.

As we got up, the inchworm slid down to the ground and humped through the grass to the trunk of the palm tree.

Maggi scooped her bag up, and we followed the mute. He seemed large for his size, like a fleshy eunuch. As I started across the street, my feet didn't seem to connect to ground. I wasn't quite sure when my foot would touch down from the curb. It touched pavement with a sudden jar. Hmm, not as far as I thought. My scalp tingled in a circle as if I were wearing a fedora hat. Who would wear a fedora in Morocco? The blare of an auto horn jarred my nerves; exhaust fumes choked up my throat. I scooted up on the curb on the opposite side. The mute boy reached out to grab me. I pushed his arm away. He wailed and flapped his arms.

"What's he saying?" I asked Maggi, thinking he was pissed at me.

"He wants to know if you're hungry," she replied.

My stomach quivered and nausea stroked my gullet. Whoa boy, I thought, and shook my head. Mint tea, yeah, with lots of sugar; that's what I needed, almost as good as Coca-Cola for settling the stomach. "How about tea?"

Maggi made a slurping sound to the mute and pretended to lift a glass to her mouth. He turned and pointed to the archway that led to the medina. We wove single file through the jostling crowd. Maggi's bag bounced from one man's djellaba to the next. I watched carefully to make sure nobody ripped it off. The stone pavement felt like warm tar, and my feet were sinking in. I marched along lifting them extra high, as if trying to stay out of quicksand. Just inside the archway, the dumb boy gestured me to a table on the sidewalk.

"Don't let on that we're stoned; somebody will take advantage," I whispered to Maggi.

"Who's stoned?"

"Good point."

A man threw open the sash of an ornate carved window on the second floor, smiled a toothless grin, and yelled down. "Oh, you, how are you? Come on up!" He pointed to a set of steep narrow wooden stairs next to the tea shop. It was the hash-cookie man. Maggi waved back enthusiastically.

Stairs, I thought, oh, shit!

"It's the cookie man. Come on, let's go!" bubbled Maggi.

"No, can't make it," I said. "I'll sit down here and guard the entrance."

"What?" asked Maggi.

"I mean, I'll catch you later down here." I flopped into a chair. The mute boy sat down in the chair opposite me and signaled the waiter. Maggi was climbing the stairs, her bag bouncing on her thigh ahead of her. A group of men at a nearby table sucked on hookahs; the gurgling soothed my nerves like a cat's purring. I wondered if they could tell I was stoned. I smiled at them meekly and waved. They smiled and waved back.

The waiter set three glasses of hot mint tea on the table. I fished about in my pocket, but the dumb boy gestured and slipped the waiter a hundred-dirham note. I reached for my glass, scorched my fingers, and yelled, "Oh, hot!"

As I shook my hand limply from the wrist, the dumb boy smiled, rocked his head, and hissed out from the depths of his gullet, "H . . . H . . . H . . . H . . . A . . . A . . . !"

"Right, H H A *W T!*" I enunciated for him.

He laughed, shook his head, and then pointed enthusiastically at a large stone tower that rose up next to the archway entrance. He pushed his chair back, stood up, gave two little hops, and gestured for me to follow him. At the tower, he showed me niches in the worn brick at the base with smooth pebbles set out in rows of ten in them. Then he craned his neck back, pointed to a blinking red light on the top of the tower, drew his finger across his throat, and made a bunch of muffled barks. We went back to the table.

The tea was cool enough to drink when we sat back down. The hookah men nodded politely, smiled kindly, and carried on with their bubbling.

Maggi came bounding down the steps and held out a large paper bag to me. "Look what he gave us, more hash cookies, for free!"

Free hash cookies? I was astonished. The poor give things away; the rich never do.

"What have you two been talking about?" she asked.

"Oh, he was just explaining to me about that tower over there. You see that red light on top? That's to keep airplanes from crashing into it and decapitating it. And the people lay out small smooth pebbles in rows on the bricks for good luck, and to keep airplanes from crashing into the medina."

"Wow," she said, squinting at the tower.

We finished our tea, and the sweet mint taste filled my stomach with a golden warmth. Cool dappled sunlight filtered down around hardwood tables and chairs spread out on the sidewalk.

"Ready to tour the medina?" asked Maggi.

"Let's go!" I said, throwing the cape back over my shoulders.

The dumb boy led us along cool cobbled streets lined with food stalls, restaurants, and stores. The smells of thyme, saffron, and cardamom filled the air. In the weavers and spinners section, large skeins of dyed wool draped across the street and the sun shone through red, yellow, and orange.

With cotton skirts rustling softly, women in veils padded past quietly, their dark kohl-lined eyes smiled at Maggi and me. Men pulled bumping carts of fruits and vegetables along the street. A man rode past on a donkey, its hooves clicking on the cobbles. The air smelled of the hay of fresh droppings. I smiled, beamed, and waved to people. My gums and the base of my teeth tingled.

"This is like paradise; everybody's friendly."

"It's because of the mute boy," replied Maggi.

He led us through an arabesque arch into the courtyard of a mosque with walls carved in ornate abstract designs. Maggi bought an English translation of the Koran and a set of prayer beads.

Instead of being the dark pagan den of despair that I had imagined, the medina was full of light and everybody seemed happy and friendly. Another of my distortions of reality was shot to hell.

We floated through the artisans' section. A man held a wooden turning bow between his toes, sawed it back and forth with his foot, and chipped away with a chisel at a spinning round of wood.

The six o'clock prayer call crackled out from minarets all around us, like quadraphonic sound out of sync. The men stopped work and quietly knelt on small mats in front of their shops and workbenches. We stood quietly as a lush, swelling peace fell over the medina. Another muezzin cry and everybody pushed themselves up and rustled back to work.

The mute boy led us back to the throbbing drums and cheering crowds of the plaza. I fished in my wallet for baksheesh, came up with a hundred-dirham note, and offered it to the boy.

He waved it off with a shake of his head, held his hand out, and gave me a thumbs-up handshake. His mouth contorted, stretched, and quivered. He set his lower teeth, and let out a long hiss of air, "f . . . f . . . f—fff—fffRREEEN DDhh." Maggi hugged him. He gave her a bear hug and a bunch of growls.

Back at the hotel, Maggi pulled out the bag of hash cookies. "Ready for supper?" she asked.

"Yeah," I replied, "I'm getting to like the taste."

We lay back on the bed, staring at the ceiling. Soon the surge of energy rippled through my body, tickled the bottom of my stomach. I began to feel queasy again. Tingling numbness swept into my scalp. Pinpoints of light exploded in my eyes. Maggi quietly lay beside me, mouth slightly open, staring intently at the ceiling.

When I glanced up, the complex pattern of rain stains and shrinkage cracks on the ceiling turned into opulent breasts. Hologram theory of reality, I thought. Take a random pattern, the brain interprets it however it wants, like connecting the dots in Sunday comics puzzles, like children making sheep and bunnies out of drifting clouds. I just happened to have breasts on the brain. As if to confirm my hypothesis, the polygonal paint cracks turned into hairy crotches and the entire ceiling became naked women dancing and undulating. I reached over and caressed Maggi on the nape of her neck. No reaction.

"Do you notice the ceiling?" I asked.

"Oh, yes, isn't it marvelous?" She replied without looking at me.

"Sure is." I slipped my hand under the top of her red blouse and caressed her breast with my fingertips.

"Maybe it's the third Bardo."

"You mean couples copulating prior to rebirth?" she asked.

"No, naked women dancing, could be paradise." I stopped caressing her breast.

"Typical male concept of paradise," she said with only the thinnest edge of good humor.

I pulled my hand out from her blouse, wrapped my arms across my chest, and buried my hands in my armpits. "Well, what do you see? Bunnies and sheep?"

"Oh, no, no sheep. It's Christ on a hill. The plains around him are covered with Assyrians."

"Assyrians? Where did they come from?"

"Syria, I think," Maggi replied. "Anyway, Christ is addressing them."

"I thought Saint Paul did that." All I could see was naked women, no matter how hard I tried. Perhaps the Assyrians were really naked women, but that didn't make any sense. Would Christ address plains of naked women?

"No, I'm sure it's Christ," she answered.

"Are you sure that they are Assyrians?" I asked.

"Positive."

I started to doubt my vision and began to feel nervous. Out of the blue I decided to grab the bull by the horns. "Do you, do you feel like making love?" I ventured.

"Not right now, not in front of Christ and all these Assyrians. Let's wait a while. Do you mind?"

"Sure, I'm in control of my situation," I lied. In fact, I was losing my grip on reality; the more I tried to control my situation the more confused I became.

I decided to deal with the women on the ceiling myself, if necessary turn them into Assyrians. I slightly crossed my eyes and concentrated on the ceiling through tunnel vision. The naked women moved closer. I relaxed my eyes and let my vision go peripheral. The ceiling retreated. Great game, I thought. I made them come closer again and then go back. Back and forth, each time closer and closer. Then with a spasm of queasiness, a churn of the stomach, the realization dawned in my brain that the ceiling would not go back. I was overtaken by the fear that I would be assimilated into a world of naked bodies and might never return to normal.

Got to stop fucking with the ceiling, I thought. I turned to examine the floor but a tingling fear in my spine told me the ceiling would come down and engulf me if I looked back up.

I decided I needed Maggi's help. "How you doing, Maggi?"

"Oh, great! Christ is so beautiful."

"I'm having a problem over here," I said.

She turned to me, full of concern. "What's happening?"

"Oh, I started making the ceiling go up and down with my mind, and now I can't get it to go back up. I am afraid I'll lose control."

She reached over and took my hand. "Don't worry; go with it, it'll be okay."

"Yeah, yeah. Right. Let go of it. I'm trying too hard to let go. I mean, I have to let go of letting go. Oh, shit! I can't go through that kind of fear again. I'm tired of being afraid of dying."

"Why? You know there's life after death."

"I'm not afraid of death," I said.

"What then?"

"The shame and humiliation that precedes it."

"Nobody wants to humiliate you."

"Yeah, yeah. You didn't grow up in Boston. I think I gotta get outta here; it's too claustrophobic. I need fresh air. I'm about to be sick!" I jumped up, put on my sandals, threw my cape over my shoulders, and bolted out the door.

Maggi shot up on the bed. "You gonna be okay?" she yelled after me. "You want me to come with you?"

I tore the door open, thundered down the stairs, and ran out to the street. Not the plaza, too much happening in that direction. I darted up a narrow cobbled alley. A network of Y-shaped glowing crotches followed me. The ceiling was chasing me, but it was fear of my own weirdness, fear of having to face the realization that I'm gay, weird, insane, whatever. I was freaking out and couldn't bear the thought that someone might see me. I needed to freak out in private.

The alley smelled of old urine; I was going to throw up. My face burned with fever. Oh, shit, here I was, just another freaked-out hippie. I wanted to scream, yell out. But what if the police came? Gotta slow down, go with it. I flopped down on the stones of the sidewalk. I started to heave, long convulsions, echoing like a sea lion bull in the alley.

Let it happen, whatever the fuck it is, let it fucking happen. Nothing could be worse than this, nothing's gonna be worse than this! Let it fucking happen! The glowing crotches turned into shining cobwebs of metallic colors sweeping down the alley. One after the other, black shiny spiders swept past me in the night. I tucked my head between my knees and wrapped up in the cape. I didn't care; it didn't matter anymore, nothing mattered.

"Hey, Boston, what's happening?"

I saw a white cape and cowboy boots. Without thinking, I looked up; I saw a crotch, big as life, calm, steady, dignified, and my hand was on it.

"What's this, what's happening?"

I was petrified with humiliation, couldn't say anything, and couldn't pull my hand off his crotch. It was as if my hand had a will of its own, totally divorced from mine.

"Oh, I get it, you like my cock. Right?"

I didn't say anything, turned my head away, but left my hand on his leg.

"Yeah, right, you and the rest of the world."

I stared straight ahead.

"Here, check it out!" he took my hand in his and pressed it into the depths of his crotch until its warmth consumed my consciousness.

"There, you see, ain't no big deal," he said in a kind, resonant tone. "It's okay. You ain't no different."

He took my hand off his leg, knelt, and cradled my hand in his. A wave of calmness and peace coursed up my arm from his hands and spread out over my body. It curled back and swelled into a great warmth in my chest. A green inchworm floated by on a silk thread undulating in the most absurdly sexual fashion, pursing its lips, and drinking Coca-Cola through a straw. All the hallucinations—the cobwebs, the inchworm—were semitransparent, two-dimensional, like black-light paintings on velvet. I recognized them as hallucinations, not really a part of reality, yet I trembled with the fear that they could become opaque and as real as the adobe walls surrounding me.

The alley brightened, and the sweeping cobwebs faded into dawning lightness. He guided me to my feet. I breathed deeply, fully, and stared at pink and green lights splashing across my brain like a memory of the single lightbulb swinging across the alley on a wire.

He put his arm around me, and we walked toward the alley entrance. I felt calm and clear, as if the cobwebs had swept the bugs and distortions out of my brain. The midnight prayer call scratched out from the mosque at the plaza.

I embraced him when we got to the street. "Thank you, brother."

"Hey, no problem, it happens all the time. You be getting back to the sister. She's lovin' you, I know, I know."

The shadows swayed easily with the lightbulb bouncing in the breeze, and the leaves of the trees glowed with spring green color.

Maggi was watching from the third-floor balcony. As I looked up, images of the moon arched across the courtyard like strobe images, a gap between each, then settled into a single position behind her. My brain seemed to present me with possible positions of the moon, then settled on one. I climbed the stairs, tired but full of energy, like after reaching the top of a mountain, and put my arm around Maggi's shoulders.

"Are you okay?" She slid her arms around my waist.

"Oh, yeah." I felt pure like clear mountain spring water.

We clung to each other, then tore off our clothes, and as we collapsed onto the bed, the ropes creaked. The touch of her skin soothed me, and as we caressed and rocked on the bark of love, the disparate levels of my consciousness merged into one being without fear or anxiety, totally consumed in passion. We held on to each other, hugged so fiercely that our bodies seemed to melt together, occupying the same space at the same time. Waves of pleasure rippled coherently from head to foot and back again. The image of her body grew at all levels within my brain, pushing aside all other levels of awareness. The echoes of ecstasy fed on each other.

The image of her inside my brain expanded until I was engulfed, like a bubble of light merged with her light. The energy curled up like a snake, struck out, and spun off into space. A white light surrounded us like the moon shooting magnesium flares into the corners of time; we floated there in the scent of time.

"Wow! Now, that's what I call making love," she said in the darkness.

"I think I lost some inhibitions." I laughed.

"I'd say so." She laughed and hugged me. We fell asleep in each other's arms.

CHAPTER EIGHT

THROUGH THE VANISHING POINT

At dead noon two days later, May 29, 1970, the cracked asphalt of the sun-baked road disappeared into mirages of silver heat waves in both directions. We had been taking the bus but decided to save time, be adventurous, and hitchhike the last leg to the ferry landing at Ceuta, Morocco.

"This is a good spot." Maggi, wearing a blue dress over black leotards, dropped her string bag at the base of a white concrete mile marker and tied a red scarf over her long, windblown hair.

"Hmm, look what I found." She pulled out the bag of hash cookies and held it under my nose. "What shall we do with them? We can't take them with us."

I didn't have the energy to deal with another hash-cookie trip. I looked around for a place to relax, perhaps to make love. The landscape was dotted with stone houses and corrals. In the distance, farmers plowed brown fields with camels. A faint breeze blowing south from the Strait of Gibraltar conjured a pungent aroma from scattered groves of eucalyptus trees. There was no place to hide or make love.

"It'll be all right; don't worry." Maggi handed me a cookie and took two herself. I lay on my black cape in a grass-lined drainage ditch at the bottom of the road embankment and chewed a cookie tentatively. I stared at the clear blue sky streaked with silver vapors rising from the earth.

Out of the silence, a donkey tied to a creosote-covered telephone pole brayed a wild alarm. A blue taxi, farmers crammed in like Keystone Cops in white turbans, rattled down the road. An ant, black and shiny like a Cadillac, crawled up my cape. Instinctively, before I realized I was doing it, I crushed the ant with my thumb. Normally, I would have swept the ant off without noticing, but I stared at the crumbled mass of body parts; the onyx beads reflected the blue sky like chrome.

Maggi knelt beside me. "Did you kill it?"

"Yes, I didn't mean to," I said. "Just instinct."

"Does it bother you?"

"It bothers me to kill without thinking. But if I thought about it, I wouldn't kill." Why was killing a harmless ant an issue? I didn't believe in reincarnation, that the ant would be reborn as somebody's mother.

I thought about Chombi the night he showed me *The Tibetan Book of the Dead*, a how-to book for reincarnation. *The Tibetan Book of the Dead* instructs that enlightenment is the goal of life on earth and we are reborn over and over until we achieve it. Enlightenment can be achieved at the moment of death by focusing on the white light and letting go of earthly attachments. It doesn't instruct us how to let go of earthly attachments, or for that matter instruct us how to recognize when we are free of them. The cruel joke of the cosmos is that the desire to reach enlightenment is in itself an earthly attachment.

"There are three stages to death," Chombi said. "First, there is a clear white light. By focusing on it and letting go of the material possessions, you go to straight to the Sunyata."

"And that is the true self, the state of being we are all trying to reach?" I asked.

"The true self is a ripple of energy on the Sunyata, the great Void. But you can't reach it by trying. Then again you can't try to not try. The only thing you can do is renounce material attachments and follow the Middle Way. The Sunyata will sweep you up when you least expect it."

"And if you don't enter the clear white light?" I asked.

"If not, you enter the second Bardo where you are tempted by benevolent spirits, great pleasure, and material wealth. Next evil spirits, drinking blood out of skulls, attack you. Snakes and monsters circle around you. There is still a chance to refocus on the light at this point. But after that, there is no hope to escape the wheel of karma. You enter the third Bardo with visions of sexual orgies and naked bodies."

"The Bridget Bardo." I couldn't stop myself. The words left my mouth before I noticed them.

Chombi looked at me sternly.

"Sorry . . ." I started to apologize, cringing before his gaze.

"Then there is purgatory," he deadpanned, "the place of perpetual bad puns." He laughed uncontrollably. "It's better to face demons drinking blood out of skulls than your puns. Ho. Ho Ho. Oh, yes, and on the forty-fourth day you are reborn."

I respected Chombi and *The Tibetan Book of the Dead*, but I was a skeptic. I did not believe anything unless I actually experienced it or could prove it scientifically, as with electromagnetic radiation, quantum mechanics, or relativity.

Up on the road a black Fiat emerged from the shimmering heat waves. Maggi stuck out her thumb, and her red scarf billowed up as the Fiat zoomed by. A VW camper sputtered down the road driven by a pasty-faced man with gray hair and gold-rimmed glasses. His wife peered through a patch of colored tourist stickers on the windshield. When they saw me, the skin on their faces tightened, turned ashen, and they stared straight ahead, concentrating gravely on the chore of driving.

After a few minutes a white Mercedes appeared on the road, like a starship downshifting out of warp drive. The driver did a quick double take when he saw Maggi. I instantly knew he was going to stop. He screeched to a dead stop in the middle of the road.

Maggi and I picked our bags up and ran the hundred yards to the car. A dark clean-cut man in a white naval uniform and captain's hat got out and opened the trunk.

He turned to us. "You are going to Tétouan?" His uniform was embroidered with ornate gold braids.

"Yes, thank you; we have to catch the ferry from Ceuta tomorrow."

We threw our bags into the immaculate upholstered trunk. His brown eyes, set deep beside his regal hawk's nose, scanned our faces and gave my black cape a once-over. His silver eyebrows twitched; I wondered if he could tell we were stoned.

"This is my wife, Maggi," I said. "And my name is Roger."

"My name is Hussein; come on, come! Get in!" He slapped the trunk shut, sprinted to the front of the car, and slid behind the driver's wheel next to a young woman in a gray veil, who turned shyly toward us.

"This is my daughter, Fatima," said Hussein.

"Hola," she said in Spanish.

"Hola," I replied.

"Como estas?" asked Maggi.

"Bien, gracias." The girl turned away and stared straight ahead.

Maggi and I slid into the plush rear seat, which smelled of new vinyl. As Hussein hit the ignition, the air conditioner clicked on, and a rush of cold, stale air filled the car. I felt relieved to be out of the hot sun. The Mercedes revved smoothly, almost imperceptibly, and as it accelerated, it seemed as if we were floating stationary, the hills and trees moving past us.

"What country are you coming from? My English is good, no?" he asked with a guttural accent.

"Yes, United States." I felt stoned and dizzy and wondered if I made sense.

"United States! Oh, beautiful country. I graduated from the School of the Americas in Georgia." Hussein cocked his head sideways and peered at me with one eye in the rearview mirror. He pushed it into third, and white guardrails flashed by in a blur. Maggi sat quietly, holding her translation of the Koran in her lap, and fingered the prayer beads around her neck.

"There are too many cars in the United States," I said. "I like Morocco. It's peaceful here." I couldn't focus on the conversation and hoped Hussein would drop it.

We sped past the dilapidated taxi full of farmers. Feeling queasy, I looked at Maggi, encouraging her to help with the conversation. She leaned forward, pushed her black hair back over her ear, and tried to focus on what Hussein was saying.

"In America," Hussein said into the rearview mirror, "nobody starves, everybody has money, not like these poor bastard farmers out here." He swept his hand across the throbbing horizon.

I remembered when Maggi and I had first met in Peru. I had wanted to photograph the slum family she lived with in a Lima Barriada. "They don't need our help; we need theirs," she had said.

True, I admired their capacity to live simply. It took incredible strength, courage, and faith to live in poverty. Although I gave lip service to renunciation of materialism, my emotional matrix demanded that I constantly prove myself through material success. I wanted the power to prove to everybody who ever told me that I wouldn't make it that they were wrong. I knew that my desire for power was ill founded, that I would never have enough power if I didn't accept myself the way I was. But the desire for power was ingrained in me at levels I didn't understand and didn't always have control over.

Maggi grimaced and replied calmly and earnestly, "Yes, America appears that way to the rest of the world, but we spend our whole lives just paying for our houses and cars."

Hussein scowled into the mirror, blinked his eyes, and studied us. He yanked the wheel and swerved round a herd of cows grazing beside the road. His daughter braced her hand against the dashboard. The car surged angrily into overdrive. I wanted to lie back, close my eyes, not deal with anything.

"Damn bastard farmers," Hussein said, "think they own the road."

"What is your work?" Maggi asked.

"I am naval attaché, secretary to the minister of defense. We are modernizing Moroccan armed forces with computers."

The road shot like a needle through the buzzing brown hills. Hussein took a sharp curve without slowing down. The Mercedes assaulted the soft shoulder, spit up gravel, and then shot onto the blacktop again. The rush of telephone poles pulled me violently toward the vanishing point on the horizon. The image of a crumpled car wrapped around a telephone pole surfaced from the depths of my mind and flashed through my consciousness.

Hussein slid the Mercedes around a pack of donkeys, glanced back, and cursed. "Donkeys not allowed on road in America, right?"

"Yes, right, against the law." I struggled to pay attention, to keep calm. But I knew I was on a bad trip.

Hussein jerked his head in irritation as he pulled out and passed the white VW camper. It drifted up and faded into the mirage at the end point of reality.

"Your countrymen," Hussein said. "See, even the old ones have money to drive around as tourists while your armed forces defend freedom in Vietnam. I like America. Everybody has two fast cars."

With annoyance and resentment, an involuntary reply came out of my mouth, as if somebody else were speaking. "Two fast cars is a greedy violation of the limitations of natural resources."

The rearview mirror vibrated as Hussien grimaced into it. He settled into the red vinyl bucket as if locking into an ejection seat, gripped the leather steering wheel with two hands, and stomped on the gas. He was getting more angry, and I was getting more confused. The car hit a hole in the pavement, bounced, and swerved. Hussein got control, lurched over the wheel; the white needle vibrated at 120 kilometers per hour. Even though he was scaring the piss out of Maggi and me, I was afraid if I asked him to slow down, it would irritate him more. I saw him in the mirror, looking back and forth between the road and me. His daughter stared stoically straight ahead.

Involuntarily the vision of a car crashing through a guardrail floated into my consciousness. Perhaps it was a response to the natural fear I was feeling, but it seemed strong and focused like a precognition. My conscious mind started to spin like the blur of gravel rushing by on the shoulder of the road. One car crash followed another. I knew I was losing control, and the more I tried to stop the images of car crashes, the more vivid they got. I felt like throwing up. Christ, that would be a mess all over Hussein's red upholstery; he'd know we were stoned for sure. I looked at Maggi.

"How are you doing?" she asked.

"I don't know, I'm confused—the speed, I see visions of car wrecks—I don't think I can talk." Now the images of car wrecks spun through my mind with the speed of the telephone poles.

"It's all right; just go with it." She reached over and took my hand.

Perhaps if I close my eyes, the images will go away. I closed my eyes. A flashing light swept over me, as if all the synapses of my brain started firing with a great white light damned up behind them. Once released into my brain, the world as I knew it would end. Oh, shit, this was worse. I opened my eyes and heard Hussein.

"You have car?" he was asking.

I realized in mid-sentence that Hussein was talking to me. "No," I replied hesitantly.

"Mercedes is the best car." Hussein seemed to flit up and down. *Why won't he just drive and leave me alone?*

"It's impossible for everybody in the world to own a Mercedes," I said.

Hussein glared into the mirror. "What did you say?"

"America"—the voice was no longer part of me and wouldn't be controlled—"is six percent of the world's population using up sixty percent of its wealth." I couldn't stop myself from talking. Maggi tightened her grip on my hand.

A deep growl came from the transmission. "America gives Morocco airplanes, guns, tanks to defend freedom!" Hussein accelerated to 185 kilometers per hour, banged the steering wheel with his fist, clenched his teeth. "What are you?" he snapped. "How you say, you are hippie. Right? You have hash with you?"

"No, we don't have hash," Maggi responded quickly and defensively.

"You go to jail for hash in my country," Hussein replied.

Yeah, yeah right. Oh, God, I thought; now I've done it. He's going to turn us in.

Hussein twisted his head around like an owl; his eyes roved up and down Maggi's legs and settled on her crotch. She gave out an involuntary shiver.

"She not your wife? Right?" Hussein asked.

"Please," Maggi said, "we want to get out of here."

"No! No! Tétouan very close now!" The Mercedes shook and quivered.

"We want to get out of here!" I repeated, my voice as sharp and strong as I could muster.

"No!" snapped Hussein. "Your woman does not have the fear of God. Does not have the shame to cover her face like Muslim woman, like my daughter here." He settled back against the headrest, his jaw tight and determined. The car sliced through the blur of hills that careened out of the vanishing point and bent around us like horses on a carousel.

"Oh, dear God." Maggi fingered her prayer beads. She would not bail me out this time.

Bastard, rich officious fucker, self-righteous arrogant asshole, pretentious, militaristic, chauvinistic bastard, I thought. I was tired of this shit.

My head flopped back against the seat, perspiration poured over my eyelids, and I looked into Maggi's eyes rimmed with red foreshadows of tears. Reality throbbed and shifted with the surging roadside. Scene after scene, each a different world, rushed out of the vanishing point. I retreated to recesses of my mind, and looked out at the shifting realities as if from inside a kaleidoscope.

Let us out of the car. Goddamn it!

My anger turned into visions of crinkling fenders, tires spinning in the air, glass splattering, broken reflectors, bent grills, warped rims, the slow drip of rusty crankcase oil.

"I am asking you one last time to stop!" My voice reverberated up the walls of a deep well tunneled through my consciousness

"Yes, we will stop, at police station in Tétuoan." Hussein smiled sardonically into the rearview mirror.

"God, I'm hallucinating that he's the devil," I turned and whispered to Maggi.

"Me, too. He's evil." She shuddered and chanted the rosary with renewed fervor.

My anger funneled into thoughts, which, more than premonitions, became realities, realities that I had no control over. It didn't matter. Things were already as bad as they could get; I couldn't make them any worse. The light exploded in my brain, slowed down, and focused against the back of Hussein's seat. I saw Hussein's head smash into the steering wheel; he turned and looked at me, blood streaming down his face. Everything happened at once, but, as if time slowed down, the events resolved into perfect clarity in my brain. I saw his daughter turn to me, her veil torn open, with shards of glass protruding from her cheeks, her soft brown eyes pleading. I struggled to control my thoughts.

As if waking from a dream, I saw Hussein's jaw snap open, the blood gone from his face. We screeched around a curve. The ant-black Fiat raced ahead, weaving around camels and donkeys. Without braking, Hussein plowed his palm into the horn button.

With an explosion of red dirt, the Fiat swerved off the shoulder and careened back toward us.

With perfect clarity, I recognized that I was causing an accident by thinking about it. I was so pissed at Hussein! What about Maggi, Hussein's daughter, and the passengers in the other car? All my life I had been taught that this was wrong, arrogant. A terror gripped me from a deeper source, a fear that I would make a mistake that would haunt my soul for eternity.

"I'm scared. I've never been more scared in my life," I whispered to Maggi, and rocked back and forth with my arms locked across my stomach. "I think I'm causing a car wreck. I don't know what is real anymore."

"Yes, you do. Just let go," Maggi said softly, and patted my arm.

"It's too late, too late."

"No . . . remember the *Book of the Dead*." She leaned her head back against the headrest and closed her eyes.

Hussein never noticed the rebound of the Fiat until the two cars collided. With a crash of metal, we veered wildly across the road. I heard a mute gasp from Maggi, and she pointed out the dust-splattered rear window, the prayer beads hanging from her index finger. The Fiat reared up in a geyser of dust like a stallion pawing the air in slow motion. It balanced on its front bumper for an eternity.

Oh, Lord! I had started this equally as much as if I had pushed Hussein's head into the windshield. What was I afraid of, to die? No, I was afraid of becoming pompous, arrogant, like those I hated.

Maggi's head whiplashed against the red vinyl seat, eyes shut in tight wrinkles. Instinct told me to focus on the far hills and clouds floating to the center of the vanishing point. My vision focused on a distant broad slab of sunshine that poured over a farmer and camel. I prayed silently, the prayer Chombi taught me, "*Whatever happens, let it happen out of love for all beings.*" Over and over I chanted. The layers of three-dimensional reality fused into a peaceful tableau, like an illuminated manuscript. My own stupidity and anger dissipated and unraveled. Why had I let Hussein push me into the abyss of anger? Time speeded up.

The black Fiat crashed down on its roof and rose again on the rear bumper, flipped again, landed on the front bumper, bounced in and out of the cloud of dust, and collapsed down onto its roof.

Hussein's shoulders rotated over the steering wheel. As the Mercedes whipped from side to side of the road, he wrestled it to a stop. For an instant after the car stopped, the pavement rushed forward and then slowly came to a halt.

Except for Hussein's daughter, we opened the car doors and jumped to the hot, still pavement. A hawk circled slowly, suspended in time as clear and thick as Lucite. The cells of my brain burst with clarity. I felt ecstatic at the realization of power. *I had foreseen the wreck. Oh, God, I caused the wreck.* I immediately wanted to brag to Maggi, but I remembered the black Fiat lying by the side of the road on its smashed-in roof. Was anybody hurt? Glass from the shattered windows splayed across the black road, reflecting shards of blue sky. The tires slowly spun and the hot engine ticked like a clock. Maggi and I ran to two men lying on the ground beside the Fiat, staring at the sky.

I knelt beside one of them, and I asked in Spanish if he was hurt.

"Quién es usted? El ángel de muerte?" he asked. *Who are you? The angel of death?*

"No, soy americano, nada mas." I'm an American, nothing more.

His eyes rolled up into his forehead; he shook his head, and muttered, *"Peor todavía!" Worse yet.*

Hussein rushed up, shouted at the two men in Arabic, and then turned to us. "Did you see how they drove? Those fools wouldn't pull over. They wrecked my front end. I'll call the Guardia and have those ignorant bastards arrested."

My stomach was quiet; everything seemed calm and dreamlike. A group of hooded farmers gathered at the side of the road like a chorus in a Greek tragedy. The white VW bus with the two Americans sputtered over the hill and pulled out to pass the wreckage. Hussein waved at them. The couple stared straight ahead and drove past. Hussein ran down the road, waving his arms after them.

"Wait, our bags!" I shouted to him.

He came back, opened the trunk of the Mercedes, and handed us our bags. He opened the passenger side door and helped his daughter out.

The dilapidated blue taxi of farmers crawled over the hill and stopped beside Hussein. He ordered the farmers out, directed his daughter to the rear seat, and climbed into the passenger seat. As the taxi sped off toward Tétouan, the farmers shuffled to the side of the road and joined those already there.

Then I got confused. Perhaps I had died in the wreck. How would I know? I retraced the path of the vehicles on the road, turned, and walked back toward the accident scene. I felt as though I were descending a staircase, and I saw Maggi at the vanishing point kneeling over one of the stunned accident victims, shielding his eyes from the sun.

"Maggi, I need to talk to you," I said.

"All right." She stood up and faced me.

"I'm confused. . . . I mean, how would we know . . . I mean . . .

how can we be sure we didn't die in the accident?" I was coming down off the effects of the hash rapidly.

"Hey, don't worry, I'm not your anima; you survived." She put her arms around me.

"Whatever, at any rate we are together. We can be sure of that." Then I thought maybe she would be mad at me. "Listen, I think I might have caused the wreck; it's the most incredible feeling."

"What do you mean?" She dropped her arms and looked at me as if I were crazy.

"I knew we would have a wreck before it happened. I wanted it to happen. I was angry at Hussein. I had the most grotesque hallucinations. I freaked out. I mean what if somebody got hurt just because I was pissed at Hussein?"

"Nobody's hurt," she said. "You're lucky."

"No," I said, "freaked out."

She slid her arms around me again. "Don't worry. Nobody's dead; nobody is even seriously hurt. You did the right thing."

"Whatever that means. I have no idea what happened." I looked down and noticed the broken glass scattered on the road.

"Listen, stop worrying! Look at it this way, Hussein let us out of the car." She laughed and kissed me.

We embraced, squashing her string bag between us. The wind blew her hair against my cheek, rustled her windbreaker.

One of the men groaned and pointed to his leg. I knelt down rolled up his pant leg to see if he had cut an artery. He winced in pain as I pulled the trouser leg over his kneecap. It was swollen, pink, and bruised. He sat up and examined his knee. Then he looked at me and made a motion in front of his mouth as if to say he was either hungry or thirsty. A small house sparrow flitted to the ground beside the Mercedes and hopped around inside looking for crumbs. I pulled the sack of cookies out of Maggi's bag and set it on the ground beside the man. He peered in, took a deep smell, broke into a smile, and elbowed the other man, who sat up and reached into the bag.

Maggi was trying to kick the pieces of broken glass off the pavement with her sandals.

"No! No! You'll cut your feet." I took my cape off and swept at the glass with the hood. As the two men munched cookies, grinning, the sparrow hopped out through the broken windshield of the Fiat, stopped in front of the men, cocked its head from side to side. Then it dove head first into the bag and flew off to a eucalyptus tree with a crumb in its beak.

"One thing's for sure," said Maggi.

"What's that?" I asked.

"We'd be better off taking the bus."

Was it a meaningless hallucination, a temporary lapse of sanity? Was I confused by the hashish? Or was it another example of the mysterious changes of awareness that happen in the face of death? Can I be sure that it really happened? Maggi was there, and she remembers the same details of the wreck that I do. The wreck really happened.

The accident in Morocco posed the same quandary as the rain dance at Woodstock or the blizzard in Tierra del Fuego. I had apparently precipitated a change of the future by willing it so. But how was it possible? Until I could answer that question, these experiences were really nothing more than coincidences and best forgotten.

The fact that I'd had three separate experiences did cut down the chances they were coincidences. I would feel better about accepting those events as true acts of telekinesis if I could perform telekinesis on demand under controlled conditions. But I couldn't under ordinary circumstances. In fact, only under unusual mental states did it happen. Why not all the time? Since they were unusual mental states, I couldn't rule out the possibility that I had been hallucinating. Of course, if I had been hallucinating, then a lot of other people had also been.

At some level *I knew* I had changed reality with thought. But not normal thought. The implications were awesome, confusing, and not a little frightening. I worried that if I discussed the implications openly, I would be viewed as insane, arrogant, or drunk with power. I needed to understand the moral implications. Although it was possible to change reality, perhaps we shouldn't do it without a broader understanding of the relationship between consciousness and the universe.

The important thing was not that I caused an accident, even if Hussein morally justified it, but rather that it demonstrated a fundamental aspect of our existence, that somehow our minds may be connected to the fabric of reality. My experiences showed me that no known forces were involved. Somehow focusing of thought facilitated changes of reality. But how?

Many hallucinatory and mood-altering drugs found in nature, some used by shamans and healers for centuries, have the capacity to alter brain chemistry. Hashish, yage, LSD, and other mind-altering drugs demonstrate the importance of these chemical messengers in the neural net and that our worldview at any given time is dependent on the chemical state of the body. The altered states of consciousness include, among others, schizophrenic hallucinations, euphoria, psychosis, and even a change in the sense of time flow itself. Even common drugs like nicotine and caffeine have a role in determining the nature of the culture in which their use is dominant. If the brain's projection of reality can be so radically changed, what is the ultimate reality?

On October 22, 1973, Candace Pert, then at Johns Hopkins University, discovered the neurotransmitter endorphin. Since then over fifty chemical neurotransmitters have been discovered, including serotonin, dopamine, acetylcholine, and norepinephrine.

There is evidence of a code of information controlling neurotransmitter activity. Neurotransmitters are assembled from amino acids using the DNA template in the cell's nucleus. The neurotransmitters are required at the nerve synapse, sometimes centimeters or, in the case of the giant squid, meters from the nucleus of the nerve cell. Every time a nerve cell fires, it uses thousands of neurotransmitter molecules, and must be replenished if continuous transmission of the nerve signal is to be maintained.

Since normal diffusion from nucleus to synapse would take too long, the neurotransmitters are carried by motor proteins running on microscopic tracks of another type of protein, *tubulin*. Motor proteins move by grasping strands of tubulin and shimmying along like a gymnast climbing a rope. Even with this facilitated transport, two days and more might be required for the neurotransmitters to move from nucleus to synapse. The neurotransmitters somehow find and mount motor proteins headed down the very track that will bring them to the terminal two days later.

At the neuron terminal, the neurotransmitters collect in *vacuoles,* empty spaces in the cell's cytoplasm. When the nerve is fired, the vacuoles break through the cell wall and spew their contents into the synapse. Some neurotransmitters are received by receptor sites on the proximal neuron, which then fires. The remaining neurotransmitters must be rapidly broken down by enzymes before the nerve fires again, which it does sometimes fifty times a second. It seems the cell must prepare for events several days in advance. The whole process is very elaborate yet must be carried out with great precision. This has urged some scientists to propose that the activity of neurotransmitters and their motor-protein transports is controlled by quantum waves or some other code of information hidden deep in cell biology.

My brief experiences with hashish and yage convinced me that our brains can change the way that they interpret the signals coming in from the world; the reality presented to our consciousness is alarmingly ephemeral. This means that our view of the world is not a complete one. I realized that the change of consciousness under psychedelics has elements in common with states of consciousness induced in the near-death state. I was eager to learn more about the nature of consciousness and its relationship to the universe. I began to doubt that there is a reality independent of consciousness.

Maggi and I returned briefly to the States and were married in San Francisco, where her sister was living. I had accepted a visiting professorship at the Physical Research Laboratory in Ahmedabad, India. Maggi was excited about the possibility of living in India, and we decided to travel overland from Spain. The Hindu holy man Guruji and scientists at the Physical Research Laboratory would help me understand the meaning of the near-death experiences.

CHAPTER NINE

THE PREDICTION

The white Tata Ambassador sedan pulled to a stop in front of our house, and Soma jumped out. He knocked on the frame of the open door, slipped out of his sandals, and stepped, barefoot, onto the polished concrete floor. Even Soma, who should have been used to the heat, had fine beads of sweat lining the shadow of his mustache.

"Soma, how are you?" Maggi stood to shake his hand.

Soma cocked his head sidewise and pressed his palms together in front of his forehead. "We don't shake hands; this is more hygienic." He bowed his head toward Maggi. "Then we say, *'namaste.'* When in India, say as the Indians say; just don't do as we do." He laughed and turned to me. "Oh, my God, are you looking yellow."

"*Namaste,* Soma, yeah, hepatitis, bilirubin of eight point two. Could be a new Guinness record." I was dead tired. Mainly I wanted to lie in bed all day and watch the wall lizards.

"I'm sure it is a record." Soma put his hand on my shoulder. "I've never seen anybody looking so yellow. You are looking more yellow than a mango!"

"That's the nicest thing anyone's said all day." I smirked sarcastically.

"And the child, how is this beautiful child you are going to have?" Soma put his hand on Maggi's stomach. "I see it is getting bigger, and his milk supply is also ready."

"I got gamma globulin as soon as Roger got yellow. We'll just have to wait and see." Maggi shouldered her net bag.

Soma draped his arm over my shoulder. "Oh, my God! Did anyone tell you that you are looking yellow?"

"Yes, I heard already." I meekly stabbed him in the ribs with my finger.

"Are you ready to face this customs inspector?" A mischievous anticipation brightened his eyes.

"I'm tired, sick," I replied. "You handle the bribing today."

"Oh, yes, something I'm good at. One of Lord Shiva's small sutras." Soma's eyes sparkled. "When bribing, stay with the bribe, become one with the bribe, feel the wondrousness."

"Don't overdo it." I picked up my leather bag.

"Me? Never. Let's go." He was into his sandals and out the door with a graceful two-step. Maggi and I struggled into our sandals and followed him to the car.

"Got your passports?" Soma slid into the front seat next to the chauffeur, who was wearing a white uniform, the letters *PRL,* for Physical Research Laboratory, embroidered in red over the breast pocket.

"Yes, they're here." I patted my leather traveling bag.

Maggi and I climbed into the backseat. It was one of the few times we'd ridden in a car since we left San Francisco over a year before. After we were married there, we drove to New York, caught a flight to Luxembourg, and traveled overland from Spain to India as an extended honeymoon.

We had traveled from Spain on buses, trains, and trucks. Even though we had very little money and traveled by the cheapest transportation, the people we met were amazed at the freedom we enjoyed traveling the world. We were a privileged class. But no one we met resented us for it.

We had arrived in Ahmedabad from Delhi on top of a train. The last floods of the summer monsoon had wiped out bridges, and the trains were held up for a week. When a lone locomotive finally pulled into the Delhi station, spouting smoke and belching steam, the whole of humanity rose up pushing and shoving. Passengers getting onto the train didn't step aside for those getting off. Men and women, carrying luggage and domestic animals, surfed in and out of the windows of the maroon-and-yellow coaches. We were pushed aside; the last seats were taken. The remaining passengers climbed on top of the coach cars. Without hesitation, Maggi climbed the nearest ladder, pack on her back.

Women in colored saris and men in white turbans sat cross-legged and straight-backed in a row fifteen-cars long. The chai-wallahs jumped from car to car, clinking their porcelain teacups. As the locomotive got up to speed, the women's saris began to ripple in the breeze and the men held on to their turbans. When the engine passed under the nozzle of a water tank, the people on forward cars turned and signaled by making pushing-down gestures with their hands. I actually didn't need to be told, but it was an unexpected kindness from people who seemingly would have pushed us in front of the train to get our seats. A low overpass or a coal chute became a common concern that led to a sense of community and camaraderie. Wave after wave of signaling hands and bobbing heads traveled the length of the train, side to side around water hoses, up and down for overpasses and coal chutes. If anyone got doused from not ducking soon enough or far enough, a great chorus of merriment and clapping of hands erupted the length of the train. Soon the train reached cruising speed and the delightfully cool breeze blew Maggi's hair around her face. She never looked more beautiful than with the tint of the setting sun on her cheeks that day.

We snaked across the Rajasthan desert, rocking on the rails and listening to the clack of the wheels. On the eastern horizon, the top slice of an enormous red moon quivered in the diminishing heat. Villages and huts under sparse groves of trees hung like mirages, poised on the cusp between the steam age and cyberspace. The smooth sea of sand stretched to the Bengali plains, and a low line of cauliflower-like clouds hugging the northern horizon sparked silently with heat lightning. I imagined I saw the snows of Everest in the distant clouds. The engine charged around a curve, and the smoke from the stack wafted down, reminding us of the campfires of nomad tribes who crossed and recrossed the desert for centuries, telling the same stories, worshipping the same legends, century after century. Now, in the dusk, the nomads sit around campfires and listen to transistor radios tuned to Bombay movie songs, high-soaring melodies of romance and heartache that must shake them to their core.

The earth turned slowly under the moon and churned up a tidal wave of ancient daydreams that swept over us. I was glad we hadn't got seats inside the train. We would be insulated from the moon's silver aura by glass windows reflecting the stuttering radiation of tungsten bulbs. As the moon rose into the smoky night, it turned from dusky red to cold white. With laserlike intensity, the moonbeams pierced us with a melancholy as sweet as opium after the bitter taste was gone.

For Maggi and me, riding on top of a train in the desert moonlight was as good as it got, better than first class. I wouldn't have exchanged places with a high-paying executive in a ritzy office for anything. On the open road we had met the blazing eyes of Rajasthan nomads, Hunza farmers, and Egyptian camel drivers and found our own humanity in the bonfires of their culture. We embroidered our souls with their memories, fearful that their spirits would be diminished by the glare of ionizing radiation that the technological monoculture was draping over the earth. Already the Indian government had rented a U.S. communications satellite and was beaming cooking classes into three thousand remote Indian villages.

Soma, searching through the first-class compartments when the train settled into the Ahmedabad station, was shocked when he heard Maggi yell down, "Up here, Soma; we came top-of-the-train class." I had first met Soma in graduate school. After he read my papers on marine hydrothermal systems, he invited me to be a visiting scientist at the Physical Research Laboratory in Ahmedabad.

He looked up startled. "Oh, my God! In all my years, you are the first distinguished scientist to arrive at the Physical Research Laboratory riding on top of the train."

Now we felt like English imperialists, cruising the country lanes of India in an air-conditioned car. As Maggi settled back into the vinyl seat, I wondered if she resented our newfound luxury. Were we selling out by riding in a chauffeured car? For my part, I thought life on the road was getting too hard, the weather too extreme, the sicknesses too debilitating. Could we impose the lifestyle on children? Could we support a family on the road?

When the air conditioner had kicked in, I was secretly glad for our comfort. I looked over at Maggi. She seemed happy and content in a way that made me wonder if she was tired of traveling also. White bullocks whizzed by on both sides of the road. The driver blew his horn wildly and swerved from side to side.

"You see"—Soma turned to offer us an explanation—"in India there is no right side of the road or wrong side of the road, like in the States. It is simple. People on foot get out of the way of people on bicycles, people on bikes get out of the way of scooters, and everybody gets out of the way of cars."

The edge of the road was ill-defined. The cracked asphalt merged into a jumble of vendors' stalls. The mango trees by the road were trimmed off twelve feet above the ground.

"Those trees are trimmed by camels. Camels don't move for anybody; they are very snobbish, you know." Soma turned his head with his nose in the air like a camel.

Two men in brown-stained dhotis struggled over a washing machine wringer. One fed sugar-cane stalks into it while the other turned the crank; juice squeezed out into a pitcher.

"Sugar-cane juice. I love that stuff!" I said.

"*Acha,* now I am seeing how you came to be such a mellow-yellow fellow," replied Soma.

"The juice comes straight out of the cane. It has to be safe."

"No, you see, it is not the sugar cane itself but the glasses it is served in. You see those fellows have only one bucket of water and a dirty rag to wipe the glasses. Now that we know, we will keep you from the sugar-cane juice." He shook his finger at me.

We turned on the main road to downtown Ahmedabad. Three men in white robes and surgical masks shuffled along the road.

"And what are those?" I quipped. "Surgeons waiting for an auto wreck? They won't have to wait long."

"No, those are Jains, a religious sect who believe all life is sacred. They wear surgical masks so they won't accidentally swallow a bug. Bad karma, you know."

Most Indian legends were beautifully depicted in a series of illustrated comic books available in English in any bookstore there, but I had never heard of Jains before.

"You must be finding Indian religious customs somewhat bizarre," said Soma.

"No, actually I find some aspects of Indian religion fascinating. For example, I read somewhere that one day in the life of Brahma, the creator, is four and a half billion years long. That makes the Hindu time scale considerably more accurate than that of the Archbishop Ussher."

"Who?" asked Soma.

"Archbishop Ussher added up all the begetting—the generations back to Adam and Eve—in the Bible and came up with four thousand three hundred years since God created the earth."

"Oh, I see," said Soma.

"There are still reactionaries who attack radiocarbon dating at science conferences. I mean, I am a Christian and I may have my doubts about a lot of science concepts, even the big bang, but I'm sure the earth is older than four thousand three hundred years. How did the Hindus come up with a time scale of billions of years?"

"I'm sure it's not a coincidence. The Rishis who composed the Vedas believed our minds contact a universal source of information. By the way, I know of an ashram to study yoga; would you like me to take you there?"

"Oh, yes, when can we go?" Maggi leaned forward with her hand on the front seat.

"I will take you if you like," said Soma. "So you are like all those other hippies just wandering hither and thither looking for enlightenment?"

"No," I replied. "I came to India for the sugar-cane juice."

"*Acha,* now it is I who feels enlightened." Soma cocked his head from side to side and smiled.

"Great."

We crossed a bridge over the Sabarmati River. Long lines of dyed saffron and ruby cloth were spread out on the banks to dry. Young boys splashed water on their buffaloes.

"By the way, Gandhi was having his ashram over there." Soma pointed to a grove of trees on the riverbank. "He started his famous salt march from there."

Yellow scooter rickshaws sputtered about like angry bumblebees. Our car screeched to a halt; the driver blew the horn fiercely and yelled out the window. A rickshaw driver yelled back, his teeth rotten and covered with the red juice from *kattha,* the Indian betel-nut version of chewing tobacco.

"What are they saying?" I asked.

"I don't speak Gujarati all that well, but our driver is saying that the rickshaw missed us by just half an inch, and the rickshaw driver is yelling back that there was plenty of room to spare and next time he'll try to come within a quarter of an inch. *Acha.*" Soma patted the driver on the shoulder and motioned for him to drive on.

The rickshaw driver looked after us with a broad red grin and spit out red juice on the pavement. The red spittle graffiti, sidewalk Rorschach tests, gave some anointed corners of the city a sickeningly sweet aroma. We drove along narrow streets crowded with people milling beneath the crooked awnings and shop signs scrawled in Gujarati.

"By the way, what are you having in these packages from the States?" asked Soma.

"Vitamins," said Maggi.

"You shipped vitamins from the States?" asked Soma in disbelief.

"Well"—I tried to push myself up from the seat but was just too tired—"we knew we might have a child."

"Oh, I see; well, the customs officer might think you want to sell them for a profit," Soma said as we pulled into the yard of the customs shed.

"Hence the bribing," I said.

The customs inspector, a Buddha of officiousness, sat behind a wooden table with uniformed assistants stationed around him. Stacks of yellowed paper tied together with ribbons lined the shelves and flapped in the breeze from an overhead fan.

"May I see your visa?" he asked. "By the way, why are you so yellow?"

I handed the passports to him. "I have hepatitis."

"You should be home in bed."

"I would like to be, but we need our household items."

"Household items, I see." He stroked his jet-black mustache. His hair was slicked back with coconut oil, and his face was pitted with smallpox scars.

"Yes, nothing but household items for our personal use," Maggi said.

"Mrs. Hart is with child?" he asked.

"Yes, our first," I replied.

"Oh, I see." He flipped through the passports and checked our visas. "You will be here three years?"

"Yes," replied Maggi.

"Well" said the inspector, "according to your declaration, you have seven hundred dollars' worth of vitamins."

"Yes." Maggi's eyes grew big and earnest. "When we left the States, I knew I wanted a family—I just want my children to be healthy—so we sent vitamins."

"You must think it is a miracle that our Indian children are being so healthy without all these vitamins," he replied with a tinge of sarcasm. Actually, I was amazed at how healthy the country Indians were.

"I assure you, these vitamins are only for their personal use," said Soma.

"Hmm." The inspector scanned our declaration. "I've never heard of this 'Pantothenic acid.' "

"Vitamin B6, it's good for the immune system," replied Maggi.

"And you are sure these are not drugs you plan to sell on the street?" He stared at me with startling officiousness.

"No," said Soma, "because they are foreigners in our land, they are unaccustomed to being vegetarians, and they felt they needed the extra vitamins."

"Those packages behind me in the loading area, they are yours?" The inspector's eyes fixed mine with fierce steadiness.

"Yes." I looked down nervously. It comforted me to see the familiar packages we had mailed a year before from San Francisco.

"There are more vitamins here than any family could use in three years," said the inspector sternly.

"Sir," Soma said, "this will fully explain the use of the vitamins." He slid a brown envelope toward the inspector.

The inspector looked at Soma, then inside the envelope. "What is this? Rupees? Two thousand rupees? Is this a bribe? I do not accept bribes!" he boomed out. The assistants and other clerks looked over in shock.

"No, no, that is a mistake. That is the money we are going to pay duty." Soma grabbed back the envelope. "I guess I left the other envelope in my office."

"I see, well, the duty on these vitamins is four thousand rupees," said the inspector. He filled in the amount on several forms and stamped them with grand authority.

"But, sir, that is more than double what we paid for them," said Maggi, tears welling in her eyes.

"That is the normal rate for items brought into the country for resale." The inspector briskly waved his arms, as if to dismiss the whole matter.

Soma counted out the money; the inspector handed him a receipt and signaled for the assistants to bring forward the packages.

"Oh, sir, one more thing," Soma said quickly. "We don't have room in the car today. Is it possible to send a car for them tomorrow?"

"Yes, certainly." He turned to me. "I wish you a speedy recovery."

"Thank you, good-bye." We hurried out and climbed into the car.

Soma turned to us in the backseat. "Well, Lord Shiva let me down. My bribing skills atrophied in the States. I'm so sorry."

"You are the worst briber I have ever seen." I wagged my finger at him. "You should be ashamed of yourself."

"Yes, I am, totally ashamed." He assumed a hangdog demeanor.

"Your problem is that you have become Americanized; you have forsaken your birthright, your cultural heritage of excellence in bribery."

"Oh, yes, you are right; please forgive me." Soma pretended to cower down in his seat and then laughed. "Oh, God, what a fiasco!"

"Okay, I'll forgive you, just once."

After Soma dropped us off at our house, I took my third shower of the day in hope of rinsing off the heat and dust. I felt helplessly tired. The doctor had told me there was no remedy for hepatitis except bed rest. In fact, I had never felt so completely helpless. Mosquitoes, some undoubtedly carrying malaria, flew freely in and out of our glassless windows. My doctor forbade me from taking quinine, and I lacked the natural immunity of the Indians. Malaria was no worse than a severe flu to them. That was all I needed, to contract malaria on top of hepatitis.

The next day I got worse and couldn't sit up in bed without exhaustion. I lay prostrate on my back. How did I get so helpless so fast? Ironically, the doctor prescribed sugar-cane juice to give me energy, since my liver had bailed out on its usual function of producing metabolic glucose from starches.

Santabehn, a lady who lived in a hut nearby, came by to help Maggi prepare the house for the new baby. Like all village women of Gujarat, she wore a ruffled hip-hugging skirt and a small, tight halter that left her midriff exposed in a sexy fashion even though she was pregnant. She wore anklets of silver bells and several dozen bracelets of ivory, family heirlooms passed down from her ancestors.

The pungent smell of cumin and coriander wafted into the

bedroom. Santabehn was cooking dal, and soon Maggi brought me supper in bed. "Santabehn made you some dal and rice. See if it's too spicy for you."

I raised myself on one elbow and dabbed at the meal with a chapati.

"Wow! Hot, I'll say, but it's okay, I can stand it, barely." I fanned my mouth and took a sip of water. After a few more bites I pushed it aside, too tired to eat.

"Santabehn's baby is due any day," said Maggi.

"What's that make, her fifth?" I asked.

"Yeah, two daughters, Veehnu and Seeta, and two boys, Arjun and Mohandas."

"How does she do it? How do any of them do it? I have trouble enough thinking of having one child."

I had qualms about being a father, especially about supporting a family. My near-death experiences had made a nine-to-five job seem meaningless, hopelessly boring. I couldn't muster the energy required to rise through the ranks of academia. It was a mean thing to do, to bring a child into the world without a steady job. Furthermore I was hesitant about passing the WASP curse of father-son jealousy on to a new generation. Even though I was aware of the grief this pattern would cause, I wasn't sure I could prevent it. My emotions had been programmed, and I couldn't change them for the world.

The next evening I lay on my mattress on the floor and watched two house sparrows flying in and out the window. We had tacked cloth over the windows to keep the sparrows out, but they had poked holes in it and constructed an elaborate nest on the wall light fixture. Indian houses are elaborate ecosystems complete with food webs, trophic levels, and feedback systems, with no attempt to exclude critters—no glass in the windows, no thresholds under the doors. The smell of chapatis toasting in the kitchen turned into the faint odor of burning twigs. Suddenly a wisp of smoke curled up from the sparrow's nest over the lightbulb. The female sparrow hopped up onto the side of the nest flapping her wings and spinning around. The male sparrow flew in the window and circled the nest crying out. The nest filled with white smoke, a small flame flickered at the base, and the sparrows flew out the window.

"Maggi!" I yelled. "Come! Quick!"

Maggi and Santabehn rushed in and looked in shock at the

flaming nest. Santabehn ran to the kitchen, returned with a broom, swatted at the flames, and knocked the flaming pieces to the floor. She beat out the flames with her broom and brushed the charred remains of the nest out the door.

"That might be a first among birds," said Maggi.

"Perhaps they thought they were moths," I replied. "That's enough excitement for one day." As night fell, gnats, moths, and other insects at the bottom of the wall-lizard food chain sacrificed themselves before the Phillips sixty-hertz altar.

Up to a dozen lizards lived in every Indian house. An elaborate system of fortune-telling has evolved based on the movement of wall lizards. It wasn't long before a wall lizard appeared in the shadows peripheral to the bulb. For fifteen or twenty minutes it didn't move. Then I lost interest and looked away. When I looked back, the lizard had moved closer to the bulb. Again I looked away for an instant, and again it moved. It only moved when I wasn't looking. It was like the game *red light green light* I played as a child.

I noticed another lizard approaching from the adjacent wall. Over the course of an hour the two lizards converged, like surreptitious statues, on the insects teeming like phytoplankton around the bulb.

With startling speed, the first lizard crossed to the adjacent wall and grabbed the tail of the intruder lizard, who immediately detached from its tail, dropped softly to the floor, and ran off. The first lizard threw its head back and swallowed the rival's tail. An interesting twist on the survival of the fittest. Evolution had selected lizards that could detach from their tails and live to grow a new one. It is hard for me to imagine the twist of random mutations and natural selection that would create the evolutionary processes for a wall lizard to regrow its tail. Either a tail detaches or it doesn't; there were no possible intermediate steps that could lead to a natural selection process.

The floors of every Indian house I visited, whether they were dirt, tile, or concrete, were meticulously cleaned with buckets of water swept under the door. It didn't matter if the house was the poorest shack in the most squalid surroundings; the floor was always immaculate.

With no thresholds to stop them, toads hunkered down, squirmed under doors, and hid out under pillows scattered on the floor. More than once they embarrassed me. Indians sit on the floor, and the proper hospitality is to offer a guest a pillow. At times I'd offer a pillow to a guest only to have several toads hop out.

One day a toad took to refreshing itself in our squat-down toilet. The flushing of the toilet became a question of ethics. Flush it down the drain, scoop it out, or wait until it climbed out on its own? What would a Jain do? It seemed unlikely the toad would climb out on its own. It was obviously enjoying itself, paddling around, and blowing bubbles. Besides the porcelain sides of the toilet were steep and slippery. I didn't feel like scooping him out. Who knew what other atrocities were down there with it. So I did what I always do when confronted with a difficult decision. I rationalized. Surely toads could survive the trip through the septic system. That's what toads do. How long could it hold its breath? I flushed the toilet. The toad spun around in the Coriolis maelstrom and bubbled out of sight and out of mind.

When I got bored studying the ecosystem on the ceiling, I read from the Upanishads, a collection of early Hindu sacred texts. Svetasvatara Upanishad: "One should know that Nature is illusion and Brahma is the illusion maker. The whole world is pervaded with beings that are part of him." Kena Upanishad: "Brahma is an uncanny something that changes its form every moment from human shape to a blade of grass." Brihadaranyaka Upanishads: "As a man acts, so does he become, as a man's desire is, so is his destiny."

One morning Maggi came into the bedroom with Santabehn. "Roger, look! Look at this." They sat down on the floor next to me. "Meet Gopal." Arjun and Mohandas, butt naked except for their ragged shirts, slid to the floor in the corner.

In her arms Santabehn cradled a tiny baby boy with a full head of rich black hair. She supported Gopal's head in the palm of her hand, his eyelids thick with kohl, the Indian equivalent of mascara, painted on by his sisters. He held on to Santabehn's finger with a firm fist, and when she wiggled her finger, he bounced and broadcast a smile full of the delight and hope of being alive.

"Oh, isn't he wonderful?" said Maggi, affectionately stroking his legs and looking at him with a love that almost made me jealous. Gopal gave a couple of playful kicks. I reached out and squeezed his toe. He looked up at Santabehn and jerked his leg away.

"Me." Maggi made a rocking motion with her arms. "Can I hold him?"

Santabehn handed Gopal to Maggi, and Maggi's face glowed like a Madonna's. She rocked him, handed him back, and said to me, "We better leave you alone so you can rest."

I collapsed back onto the mattress and relapsed into the study of the ceiling, so thinly whitewashed that the underlying concrete showed through.

Like our neighbors, Maggi and I slept on the roof in the dry season. Our house was on the outskirts of town. To the east was Ahmedabad, to the west the Gujarat Desert, a flat expanse of sand that sprouted thin grass in the monsoon season. We watched the sunset on the desert from the roof of our house. Every evening, thousands of crows formed a flock and flew directly over us, silent as a prayer except for the sound of their wings rustling like nuns' habits. During the day I rarely saw more than one crow at a time, usually flapping around cocking its head looking for scraps of food by the side of the road. Somehow as the heat of the day wore off, they organized themselves into a riverlike current, flew out of the sunset, and whispered over the roof of our house. It happened every day.

Maggi and I lay on our mats staring up at the crows.

"How's the baby?" I leaned over and put my head against her stomach.

"Fine."

"Are you planning to nurse?"

"Of course; bonding, you know," she replied.

Like death and tears, breast-feeding was taboo in my New England culture. My mother hadn't breast-fed me, and during my birth she had probably been under spinal block and perhaps chloroform. Birth was treated as a pathogenic disease, and the newborn as an object that had emerged from the dirtiest part of the body covered with vernix, blood, and meconium. Who would want to dwell on this lowly, gruesome origin? Who could overcome the taboo? I was probably whisked away at birth, washed, and placed in a nursery with dozens of other screaming and frightened infants. I was allowed to be with my mother for brief intervals until we left the hospital five or six days later. At checkout time I was probably presented to her like a gift-wrapped Christmas present. Like other children reared in the States, once I was taken home, I was locked alone in my own bedroom, never sure what I did to deserve such loneliness.

Maggi was my best experience and only role model for raising a child. Like many American and European mothers in the early 1970s, she was helping to overthrow the culture of male-dominated childbirth in hospitals. Women were fighting for more humane conditions, reverting to home births, isolated and sometimes afraid but willing to pay the price to avoid tongs, bright lighting, and surgical trays as the infant's first bed. We were lucky because those conditions did not exist in India. Instead, there was a network of birthing houses, where the whole family could camp out and attend the birth.

"Babies in the States are bottle-fed," I said.

"Maybe that's why so many are hyperactive."

"I thought that was from watching TV."

"That too . . . Breast-feeding is important for nutrition," she added.

"Maybe we don't need all those vitamins, then."

Maggi rolled over, grabbed my shirt, buried her head on it, and started sobbing. "I know. I know. I was just scared. I just want our baby to be healthy."

"Oh, it's okay." I put my arm around her, rubbed her back. "It's okay. I understand. The baby will be just fine. It was money well spent. Maybe the vitamins will protect the child from hepatitis."

Maggi lifted her face to me. "I'm glad we are going to have the baby in India. They have such a sensible attitude toward birth. No doctors breathing down your neck with tongs and cesarean knives." She folded her hands over her belly. "Birth couldn't be worse in developed countries if they tried. Babies are delivered in cold stainless-steel rooms filled with gloved and masked strangers."

"Yes," I agreed. "I like the birthing-house idea. I like being part of the birth." Certainly there had been no possibility of my father being present at my own birth. I hoped that I could be a better parent than my parents. My parents provided me with material security, but at times that did not seem as important as love.

"Do you think our child will be different? I mean, most kids ruin their parents' lives, demanding attention, squabbling all the time."

"That's because they're starved for love. Look at the women in Santabehn's village. They share their lives with their kids, give them unconditional love, and they're not brats."

It was true. Even though we had great riches in suburban New England, our parents tried to keep us from bothering them. In contrast, I had seen children living in utter poverty who showed love and respect for their parents and readily helped with chores. Maybe Maggi was right; maybe my fears were only a projection of my own upbringing. Then I had second thoughts.

"Maybe those kids are well-behaved because they are petrified of their parents."

"Surely not all of them are beaten every day. It seems more likely they feel secure because they feel loved. They are given to the mother to breast-feed as soon as they are born, and the child often sleeps with the whole family, in a single room. In our culture the child is locked in a room with a Mickey Mouse night-light as its only company."

She stopped, a distant look on her face. "You know the Indians have a good system. They spend the main part of their life as householders, and when they become old, they wander around as sadhus looking for enlightenment."

I wasn't ready for the householder stage, but I was interested in the sadhu concept of enlightenment. What could it tell me about my near-death experiences?

At Maggi's request I brought home some books on the birth process from the excellent library at PRL. The more I read, the more I was convinced that the physical process of birth is impressed on a metaphysical substrate. Hidden intelligence seems to guide the process.

For example, three million or so single-celled sperm enter the vagina and swim toward the eggs in the fallopian tube using the vigorous, whiplike motion of their tails, wave after wave passing down the tube. Yet the sperm possess no brain or nervous system to coordinate the swimming motion.

No less amazing is the intelligence of the egg. The woman is born with two million pre-eggs—developed in utero when her mother was three months pregnant—waiting in the ovarian wings, but she loses many during ovulation, until at around the age of fifty they are all used up and ovulation stops.

During ovulation, the egg is released along with a cloud of hormones, including a chemical attractant, a perfume, that lures the sperm up the correct fallopian tube to the ovulated egg. Cilia lining the fallopian tube speedily brush the perfumed come-on ahead of the egg. How does the sperm change direction and follow the hormone into the correct fallopian tube? Does it twist its head? Change the motion of its tail? How does it coordinate those changes without a brain or nervous system?

Most sperm die before they reach the egg. The few hundred that reach the egg batter it and release protein enzymes that strip the egg of its protective coating. One sperm enters. A special protein key on the sperm's head fits a protein lock on the egg; sperm from other species can't enter. The egg and sperm combine into one cell with the "double yoke" of two nuclei.

But all that has gone on before pales in comparison to the gene swapping that takes place next. The sperm and the egg each duplicate their twenty three chromosomes. The membranes surrounding the two nuclei disintegrate. Motor proteins, free-moving muscles no larger than a single molecule, manipulate the chromosomes on tracks of a protein, *tubulin,* arranged in fibers that radiate from opposite poles of the cell. These spindle fibers and the motor proteins work together to line up the chromosome pairs from the mother and father on opposite sides of the equator of the egg. It is unknown how the spindle fibers locate the middle of the cell. The mother's genes lock with the father's genes so that they precisely match up, gene to gene. How the corresponding genes recognize each other and manipulate into the proper locked position is also unknown.

In a process called crossing over, genes are exchanged between the mother's and the father's side of the cell. The parent cells unselfishly give up genes to provide diversity for species survival.

In a miraculous process, the genes separate into two cells. Motor proteins, attached to each chromosome, grasp with molecular hands the nearest spindle fiber. Tugging, straining, and releasing, extending and grasping again, the motor proteins pull apart the two sets of genes in opposite directions. Five to ten minutes pass as the motor proteins reel in their cargo through the ten or so microns to opposite poles of the egg. This coordinated dance of molecules can take place in vitro, outside the mother. Thus, we know there is no guidance from the mother. What is cueing this dance of molecules?

New nuclear membranes form, individually packaging the two sets of genetic material. A ring of the muscle protein, *actin,* miraculously forms at the egg's equator. A motor protein, *myosin,* equipped with hands that swing forward with a powerful grip, grabs the actin. By bending its "wrist," the myosin pulls the actin cable tighter until the cell is cut into two. The first cell division is complete.

By the second day three cell divisions produce eight cells with no increase in size. From eight identical cells clustered together in the shape of a mulberry and no bigger than a period at the end of a sentence comes a human or even a blue whale.

When eight cells divide to sixteen, the organelles within each cell are pulled toward the exterior of the eight-celled ball. Then in near-perfect simultaneity the eight cells divide. The inner cells become the fetus, the outer cells the placenta. The embryo appears as a foreign invader to the mother's immune system, and the placenta helps protect the embryo.

By day four, the embryo, consisting of thirty-two cells, completes its journey down the fallopian tube, enters the uterus, and releases, a hormone, *human chorionic gonadotropin*, that terminates the menstrual cycle. This cellular juggling act is programmed by genes that stay with adults through life. Somehow they are cued to turn on only at this stage.

After seven days the embryo releases enzymes that penetrate part of the uterus and mesh with the uterine wall. The outer cells do away with separating membranes, and the placenta projects capillaries into the uterine wall, where they contact the mother's capillaries and begin exchanging oxygen, nutrients, and waste across vessel membranes. How do these capillaries locate each other to link up? It is a lesson plan in chemical engineering. The entire mass is a hundred cells, a fluid-filled sphere, but still the size of a period.

The embryo is now supplied with nutrients from the mother, and cell division begins in earnest starting with DNA replication. Every human cell contains about two meters of DNA, crammed into the nucleus of the cell. Each thread of DNA is about three inches long; all the DNA in the human body would stretch to the sun and back. This is the wonder of being human. The nucleus, about one-tenth the volume of the total cell, is perforated by thousands of porelike protein gates. The pores keep DNA in the nucleus while allowing ready access and egress for a host of other molecules used in the replication process.

At the end of the second week the dormant inner cells of the embryo awaken and form a longitudinal axis. As the cells divide, some of the daughter cells discover or interpret how and where to become the cells of a heart, a liver, or perhaps a nose or toe. Originally identical cells become very different parts of the body. The knowledgeable differentiation among the newly forming cells is orchestrated by concentration gradients of specific molecules within the cluster of cells. How do these chemicals depict the relative future positions of organs within the fetus? The fetus appears to be following a plan not provided by nerves or random diffusion. A week later the heartbeat starts. The embryo is about a quarter of an inch long.

The cells have become full-fledged protein factories. A protein is a string of several hundred amino acids, molecules having twenty or so atoms. Individual amino acids are brought into the cell and strung into specifically ordered chains, snipped, respliced, and then shaped into the structures of life. Every cell in the human body is making two thousand proteins every second from a combination of three hundred to over a thousand amino acids. All told, each of us is organizing 150 billion billion amino acids into 100 million billion carefully constructed chains of proteins every second. Many of these are transported to specific exit portals by motor proteins. Is this a mindless process of molecules accidentally bumping into each other?

Picture a three-dimensional intersection of several major, multilane highways—crossovers, on- and off-ramps, an interlacing of cloverleafs, one above the other, traffic moving in all directions. Reduce it to a millionth of a meter and repeat it ten thousand times in a single cell. The roadways in the cell are fibers of tubulin along which motor proteins transport newly formed molecules from point of production to point of need.

Basic cell structure is the same throughout the biosphere, from mushrooms to humans. Thousands of receptor and transporter proteins penetrate the cell wall, determining what can and cannot pass through. Proteins do not form outside of a living cell, even in a laboratory test tube, because the plan of controlled transport is missing.

At week six, eyes are clearly visible, arms and legs sprout with digits fully joined by webbing. During the next two weeks, the cells of the web die and the fingers separate.

By the ninth week the arms and legs, hands and feet, and most of the body look clearly human.

The brain begins to enlarge. The deep structures form first. The brain stem controls breathing, heartbeat, and smooth-muscle contraction as well as the twelve muscles that direct the motion of the eye in a synchronized fashion. The limbic system controls the emotions, wakefulness, and attention. The cerebellum automatically controls the force with which we place a foot on the ground. If it miscalculates, we receive a jolt. When we rotate the head, the scene apparently moves, but the cerebellum lets the visual cortex know that the head, not the scene, has moved. The earth is rotating at a thousand miles per hour, and we don't feel a bit of it because of the cerebellum. In a sense, the theory of relativity is an expression of how the cerebellum controls our perception of reality.

The cerebral cortex, the part of the brain used for thinking, is visible at this stage. The human cerebral cortex has ten times more neurons than the chimpanzee's, even though there is less than one percent difference in the genome. The last to form are the frontal lobes of the cerebral cortex, and they continue to develop throughout childhood. The frontal lobes are associated with conscious thought, associative reasoning, speech, and control of emotional impulses. The frontal lobes take data from the visual cortex and decide what to forward to the conscious mind. Self-awareness emerges at this stage. Messages are relayed across the brain and from the brain to various parts of the body.

New nerves form at the rate of four to five thousand each second. By birth there are one hundred billion neurons. The adult brain has one neuron for every star in the galaxy. Each neuron connects with as many as one hundred thousand dendrites of other neurons, over a million billion connections in all. A complex action—such as simultaneous wiggling of toes and fingers, moving of arms and head, and reading—can involve the combined action of a million links, each cycling five times a second. Many neural connections are hardwired, set in place by the genetic instructions of the cell. But many are not.

Our brains have two types of neural circuitry. One type is like the fixed read-only memory on a hard disk, the hardwiring. The other is like a self-correcting software program that is malleable and develops in response to experience. The brain hunts for certain logical relationships, and when they are not there, it makes deductions drawn from its personal history. It fills in the blind spot, where the optical nerves pass through the retina. It fills in the space in an iron bar, even though it is less than one percent solid. Matter is the brain's conscious expression of information.

Our brain gives us the illusion that we feel our toes in our shoes, but all physical sensations are registered in the head—a triumph of ideas over physical reality.

My reading made it clear that the human brain is the most complex structure we have yet encountered in our universe. If the conscious mind is not mystical, it is at least metaphysical.

Several weeks later, after I had recovered from hepatitis, Soma picked us up in the PRL car and we drove toward the Sabarmati River. We got stuck behind a large two-wheeled cart being hauled by two massive white bullocks with six-foot horns. The body of a gray cow was limp in the cart. Our driver honked his horn impatiently. Without pulling over or speeding up, the cart lumbered forward as if we didn't exist.

"So this is gridlock in India," I said.

"Untouchables," said Soma. The adobe walls along the road were covered neatly with cow pies. Each one with the handprint at the same angle. "The untouchables also collect cow pies," said Soma, "and dry them for fuel. Actually, a very efficient system. In India everything is used; you won't find a scrap of paper on the ground, a few fresh cow pies maybe, but not a scrap of paper."

"I thought Gandhi did away with the caste system and the untouchable stigma." Maggi looked into the cart as we passed.

"Technically, yes, those fellows are no longer untouchables. They are not considered lower than anybody else, but there is still an economic caste system. Those fellows are the only ones who will butcher a dead cow."

"What do they do with the meat?" I asked. "Hindus are vegetarians."

"That's correct; they sell it to the Sikhs and the Muslims, who will eat anything, even Hindus, or so I was told as a child." He laughed then pointed to a man squatting in front of a wooden crate. "You see that shoe repairman? Every place he walks he is looking for scraps of leather or vinyl. He picks up the tiniest nail and straightens it out."

A man and a woman pulled a wooden cart with huge balloon tires piled high with crates. A family stretched out on one cot under a mango tree.

"Are those people living on the street?" I turned, looking out the window as we sped past.

"Oh, yes, millions of people live on the streets in India," replied Soma. "It is not considered shameful."

"But how can they deal with the lack of security?" I turned back to Soma.

"They have their faith," replied Soma. "Renunciation of material possessions is a high goal in India. Just the opposite of the States, where poverty is a source of shame and humiliation, and a sign of defeat."

"Why doesn't the government impose birth control?" I asked.

"You see, it is like this," replied Soma. "The government has tried, Indira Gandhi tried, but it's useless. To the Indian, children are sacred. Children are security. You see, we don't have Social Security or Medicare, so it is essential for every Indian married couple to have at least one male child to support them in old age."

"But if there were fewer of them, would they all be starving to death?" I already knew the answer, but I wanted to hear Soma's version.

"Very few actually starve to death; they are poor and probably hungry, but they do not starve to death. In fact, in all of India with over twice the population, we eat less food than all of the U.S. You use more fertilizer on your lawns and golf courses than we use in all of Indian agriculture."

"I wonder if they are poor because they have faith, or if they have faith because they are poor," I said.

"I suppose both, depending how you look at it. You see Hinduism is free of structure. You can do whatever you like. Worship a monkey or an elephant. Nobody is going to say you can't. I myself have seen levitations. A young boy meditating under a tree rose off the ground."

"And you call yourself a scientist. Was it a trick?" I asked. "You know, smoke and mirrors."

"It doesn't matter. Even if it was smoke and mirrors, it was fantastic. You see, we believe the mind is a hot line to God, only there is static on the line. When you get rid of the static, the voice of God comes through. It doesn't matter what you meditate on—Brahma, Mohammed, Christ, an elephant, or an ant. The important thing is to quiet the mind and hear God; well, you don't really hear him so much as feel him. Everything else is a metaphor."

"So you are saying our mind extends beyond our brain?"

"Of course. The average human mind contains ten million times more information than all the computers in existence."

"You're assuming the mind is digital. Perhaps it's analog and there is subtle signal modulation going on."

"Even so, do you think all that information is contained in the physiological brain? Experiments on rats show that after the brain is cut to one-tenth its normal size, it still contains all that information."

"What about this guru we are seeing today. What does he worship?"

"You'll have to find that out for yourself," replied Soma.

The car pulled in front of an ochre-colored house framed by locust trees. We entered through an airy front porch with maroon window hangings flapping in the breeze. We left our sandals with several dozen other pairs and entered a high-ceilinged room with a slowly rotating fan. A dozen students in white shorts practiced yoga postures.

"Guruji," said Soma, "I've brought you two friends from the States."

"Come in!" A pudgy man with snow-white hair and beard and a fantastically smooth brow beckoned us to enter.

Soma knelt in front of him; Guruji put his hand on his head and said, "Narayan."

Maggi and I did the same.

"What does 'Narayan' mean?" asked Maggi.

"It is one of the names of Vishnu," said Guruji. "It means, 'God bless you.' "

The phone rang in the next room. Guruji got up and answered it. "Guruji, here," he said, and he carried on in Gujarati.

Maggi smiled. "It's as if someone picked up the phone and said, 'God speaking.' "

"It might be," said Soma.

"He is like Santa Claus," said Maggi. "I just love him."

"Let me tell you, he is wonderful," said Soma. "He was in an auto wreck. I visited him in the hospital. When I asked him how he was doing, he replied that his body was doing terrible but that he was fine. That is what yoga is about, becoming detached from the body and the whole material plane."

Guruji came back and sat down on his cushion, smiled at us, and said, "So how shall I be helping you?"

"My friends would like to study yoga," replied Soma.

"*Acha,* very good. Come tomorrow morning six o'clock and bring towels to practice on."

Several months later, Guruji sat beneath the full-length portrait of himself, watching over the yoga class with an amused benevolence, stroking his long white beard, greeting a steady stream of visitors, and giving out blessings.

Maggi and I had been attending yoga class every morning, practicing *asanas,* yogic postures, thinking that at some point Guruji would teach us meditation. I told him about my experiences on Mount Everest, Tierra del Fuego, and in Morocco. I hoped he would shed light on them. I was eager to get on the road to enlightenment.

We decided to approach him. "Guruji, can you teach us meditation?" I asked.

His face lit up. "*Acha,* come with me." He led us to a small darkened room lit by one candle on a table in front of icons of Shiva and Krishna. The room smelled sweet from incense and flowers, and perhaps because it was dark, it felt refreshingly cool. He motioned for us to sit on the thick carpet.

"The goal of meditation is to get the normal activity of the mind to turn into silence but without coercing it. It is the act of not doing, but if you try to not do it, you will fail. Don't count. Don't control breaths. Make no effort. Do not resist falling asleep. Do not resist anything."

"How long will it take for us to reach enlightenment?" I asked.

"It's impossible to say. Most likely you will reach it, gradually, in steps, how do you say, in epiphanies, small awakenings, without being aware of it. In fact, if you are aware of it, you have lost it."

"It sounds difficult," I said.

"No, it is easy." The way Guruji smiled made it clear he'd had firsthand experience. He was the role model. "First you chant for a time. The vibrations of the sound will heal your body, tune your cells, and help connect your body with your mind." He sat in the lotus position and chanted long refrains of "Aum" in deep resonant tones: "Like this. *Auummm. Auumm.* This is the first word, the most basic sound from which all others are derived, the sound equivalent of the cosmic energy."

Maggi and I tried the chant, Guruji looking us each in the eye with kind concern. "Wow, I feel like I've been waiting all my life for this," said Maggi, a rosey glow of euphoria spreading across her face.

Guruiji watched us with an expression of care and empathy. He corrected us a few times and then said, "Now, for the last stage you must have a mantra. Focus on this mantra and it will lead you like a seed to the universal mind." He whispered the Hindi word *hrim* to each of us.

"We heard we're supposed to keep mantras secret," said Maggi.

"Oh, yes, yes." He laughed. "I've heard that too, but no, no, there's no need to keep it secret. You see, I believe meditation is like a gift. The mantra is my gift to you."

"What does it do?" I asked.

"The mantra is like a taxicab that takes you to pure consciousness; then you lose the mantra."

"And the mantra will lead us to enlightenment?" I brushed the carpet in front of me with my hand.

"It will help you, but it is very difficult for Westerners to reach enlightenment. You try too hard. Enlightenment requires that you give up seeking enlightenment as a goal or achievement. Finding enlightenment is not a competition or a piece of merchandise."

"Can you lead us to enlightenment?" asked Maggi.

"No, I cannot," replied Guruji, shaking his head. "Nobody can do that for you, nor should they pretend that they can. You must do it for yourself."

I was devastated. We had come all the way to India to find enlightenment and now Guruji was telling us that he couldn't give it to us. Secretly, I wished I could brag to my family and friends that I had found enlightenment in India. It might earn me some respect, justify my existence. On the other hand, I respected his honesty, however brutal. In fact, I despised the "Bliss Bunnies," as Maggi called them, that followed gurus around, acting enlightened. They were being pious to satisfy their egos. If the near-death experiences had taught me anything, it was that disempowering the ego was the first step.

"It is like this," he replied. "The goal of life should be to find the cosmic consciousness that exists in every experience."

"Oh, my goal has been to get even with everybody who said I couldn't make it," I said.

"True, that's most people's goal. But it is a waste of time. The goal will never be reached. You are the only one who can decide whether you make it or not. The people who told you that you wouldn't make it are poking your eye out because they feel insecure themselves. Like Gandhiji said, an eye for an eye makes the whole world blind."

"Can you tell us what it is like?" I asked.

"In the Bhagavad Gita, Lord Krishna tells the warrior Arjuna that every person shelters a 'Dweller in the Body' who is wholly unlike the isolated, vulnerable ego. Each of us has an individual spirit that undergoes unique experiences in life—this is called the *jiva* in Sanskrit, similar to what you call the soul."

"So when I separated from my body on Mount Everest, it was the jiva?" I asked.

"Yes," he replied. "You awoke to realize you have a second body, a mental body composed of nothing but energy, formulated by nothing but pure awareness. Death is like walking into your mind."

"This second body," asked Maggi, "what is it like?"

"The second body can be modified into whatever form is desired."

"For example?"

"For example, a shapeless cloud of energy, a mist of colors, a pattern of waves, or a ball of light."

"So this second body is the same as the mind?" I asked

"Yes," he replied.

"And the mind survives after death?" I asked.

"Yes, and the mind creates our physical bodies. We are what we think we are. The human mind is a template that guides the form and structure of the physical body. We are like interference patterns from two waves of energy, one direct from the divine that we shape with our thoughts, and the other indirect from the divine via our environment."

My mind had drifted off and was studying the statue of Lord Shiva in cosmic dance, balanced on one foot with the other bent up.

"So, do you see what I mean about the second body?" Guruji smiled, as warm as the sun.

"Oh, yeah, sure. So my fall on Everest was like enlightenment?" I looked from Shiva to him.

"Yes, it is like that. Sometimes a man must die to truly live," replied Guruji, smiling. "If a tiger attacks a man, at that moment he will be enlightened, but only temporarily, especially if he is eaten by the tiger. For example, you remember what the Void is like from your near-death experiences, even though you are not in touch with it all the time. In this way the Void is slowly becoming incarnate in materialism. We don't know why; we just know that it is so; it is fundamental."

"The Void?" I remembered that was a term Chombi had used.

"Well, it's not really a void, in the sense that it is empty. The Void, or Brahma, is formless, but is the birthplace of all forms, the word of God, if you will—a seamless whole that controls all the shapes and forms, the realm underlying the world of objective phenomena. Reality appears out of it and then enfolds back into it in an endless flux. Like making clay statues out of mud. All potentialities and possibilities are there. The Void can be activated anywhere, its effects do not diminish with distance, and it exerts a subtle push on the surface of existence, like water pushing on things immersed in it. All entities are ripples on the sea of energy in the Void, just as waves undulate on the sea."

"If it is full of energy, why is it called the Void?" asked Maggi.

"It is called a void," continued Guruji, "because it cannot be measured or perceived directly. It is subtler than matter, more fundamental. Matter emerges from patterns of energy in the Void."

"So enlightenment is tuning in to the Void?" I asked.

"Yes, enlightenment is merely the attuning of our central nervous system to states of consciousness that always existed in us but have been blocked by our mental conditioning," said Guruji.

"Then most of the time we are lost in false mental conditioning," I said. "And during my fall on Everest, I gave up the mental conditioning and experienced the true reality directly."

"Yes, but it doesn't have to be a near-death experience. Anything that subordinates the ego—like suffering, for example—works," replied Guruji.

"Suffering brings enlightenment?" I asked.

"Yes, suffering keeps one from falling into the trap of false enlightenment, that is, not truly letting go of the illusion created by the ego. It forces you to give up your separateness and experience the cosmic reality directly. In the West, you have a worldview that fights to conquer and deny suffering."

" 'Trap'?" I asked.

"Yes, the trap of complacency and beatific feelings that are delusions. Here we worship Lord Shiva, the destroyer. He pulls us in all directions until we are pulled into the directionless."

"So during the fall, when I gave up all hope of survival, I was forced to give up the illusion."

"Correct."

"So that is why in India you place no importance on material things. They are a distraction," I responded.

"Yes, the Buddha taught us that material possessions are a distraction, an illusion of success and happiness that keeps us from realizing the true meaning of existence."

"And suffering takes the power from the ego," said Maggi.

"Even more than that, the ego creates suffering. All suffering lies in the superstition that you live in the world, when, in fact, the world lives in you."

"Once I was suffering because Maggi was out with another man," I said. "It was like an emotional death; I mean, I truly felt my life was meaningless without her."

"Yes," replied Guruji, "sometimes an emotional death is as strong an experience as a physical death."

"Well, the strange thing is that I had a strong feeling, a precognition, that I would bump into her; then I did."

"Yes, in a dark, crowded movie theater. It was amazing," said Maggi. "It couldn't have been a coincidence."

"Very few things are," replied Guruji. "Coincidences are one way of communicating with the reality of the Void."

"But all this is unscientific, unproved," I said.

"What is scientific proof?" asked Guruji.

"Observation, deduction, and hypotheses confirmed by predictions," I replied.

"Would you believe me if I made a prediction for you?" replied Guruji, tilting his head, his eyes gleaming.

"Well, probably—depends on the prediction," I replied.

"You must assure me that if my prediction comes true, you will give consideration to my views," he replied.

"Okay," I replied.

"Okay." Guruji's smile turned into a long steady stare. His eyes, deep and silent, studied me, as if looking through me. The hairs on the nape of my neck bristled. I turned around and looked behind me. His forehead brightened and he said, "I predict you will have a son born on May twenty-ninth."

That's not a great prediction, I thought to myself. Maggi was already two months pregnant, and it was a fifty-fifty chance that it would be a boy.

Guruji's white beard hung off his broad smile like a temple curtain. He shrugged, turned his palms to the heavens, and said. "Well?"

"Yeah, sure, why not?" I looked down at the rug.

"A boy!" said Maggi. "That's wonderful."

"Oh, yes, you are blessed, Mrs. Hart," replied Guruji. "In India we believe a male child is the best gift you can have. In fact, that is why we have so many children; always we are looking to have more boys."

"Girls are good too," replied Maggi.

"Oh, yes, they are divine," replied Guruji. "But since we have no retirement plans, no Social Security, for when we get old, it is the responsibility of the male children to take care of their parents in old age."

I rode to work at the Physical Research Laboratory on a bicycle along desert lanes lined with prickly pear cactus. Occasionally, cactus debris with sharp needles littering the pathway punctured my tires, especially in the intense heat when the rubber thinned out. Since half the population of India rode bicycles, there was always plenty of work for the bicycle repairmen who crouched in the shade of wooden crates every quarter of a mile or so.

My office at the Physical Research Laboratory was in a newly constructed annex. It seemed perpetually unfinished; the last bits of splattered paint and cement were never cleaned off the windows. A troop of langmuir monkeys would sit on the windowsill outside my office, then jump about and pull on each other's tails. With uncanny timing they could devastate a garden just the day before it was due to be harvested by the owner.

Soma had just set up his alpha-particle-counting laboratory. He analyzed beryllium-10 and lead-210 and dated the age of sediments in the fans off the Indus and Ganges rivers. In addition he had used carbon-14 and silica-32 to determine the age of the groundwater under Gujarat. It turned out to be over six thousand years old. As I sat in my office, I could hear Soma's sari-clad lab assistants sing Bombay movie tunes as they did their work. Soma sang along with them. This was unlike any laboratory where I had worked in the States.

My office mate, Goswami, was a particle physicist. He studied the particles in the high-energy cosmic rays bombarding earth's upper atmosphere. He sent blocks of Lucite aloft in high-altitude balloons; the cosmic rays penetrated the Lucite, breaking bonds between molecules. Goswami then etched the Lucite in acid, and the tracks of the cosmic rays were revealed under a microscope. He determined the mass and velocity of the particles by the length and curvature of their tracks.

I was sitting at my desk working on a paper for the Deep Sea Drilling Project, and he was working at a fume hood on the inner wall of the office.

I looked up at him. "Where do the cosmic rays come from?"

He turned to me, his safety goggles sticking out past his black beard. He said, "First of all they are not really rays. They are actually charged particles traveling at immense speeds. Some, such as the nuclei of iron, were probably thrown out during supernova explosions. On the other hand, some of the most energetic cosmic rays appear to come from beyond our galaxy. Perhaps they are accelerated by magnetic forces in the galactic nucleus. They hit earth's upper atmosphere with the energy of a baseball thrown at high speed—a single 'cosmic ray' can light up a lightbulb. They bombard earth and help generate an electric charge in our atmosphere—there is a two hundred volt charge between our head and feet, but we don't notice it because we are adapted to it. Isn't it interesting that some primary cosmic rays appear to be neutrons?" Dressed in a white laboratory coat, Goswami turned to the fume hood.

"Why's that?" I asked.

"Normally, by themselves, neutrons decay. In other words, outside of the atomic nucleus, a neutron transforms into an electron and proton within fifteen minutes." Goswami pulled a Lucite block out of a beaker with a pair of tongs.

"Fifteen minutes, that's hardly time enough to travel from the nearest planets."

"That's what's interesting. There is no possible source of energy in our solar system to produce them. So either they are secondary—produced by the impact of other cosmic rays with atoms in our atmosphere—or they are from beyond our solar system."

"How's that possible?" He dropped the Lucite in a beaker of water.

"At least some of them are coming from the direction of an X-ray star in the constellation Cygnus. The decay of the neutron is precise, like a clock. Nature's most accurate clocks, atomic clocks such as the cesium clock, are based on the rate of radioactive decay. We are certain that a neutron always decays in fifteen minutes. So these neutrons reach earth in fifteen minutes from sources that are thousands of light-years away."

"How is that possible?" I ask.

"Einstein's theory of relativity predicts that clocks slow down as they approach the speed of light. The neutrons in cosmic rays are traveling close enough to the speed of light that the neutron clock, the rate of decay, has slowed down." When Goswami removed his goggles, his eyes sparkled with the light of a philosopher.

"Fascinating. So the fact that there might be extraterrestrial neutrons in cosmic rays supports Einstein's theory of relativity."

"Exactly. At close to the speed of light, fifteen minutes translates to thirty thousand years—more than enough time to travel from the postulated black hole in the X-ray star Cygnus X-2."

" 'Black hole,' what's that?" This was the first time I had heard of the term introduced by physicist John Wheeler in the late 1960s. The physics group at PRL was active in the theoretical study of the superdense matter found in neutron stars and black holes.

"A black hole is the massive remains of a supernova explosion. If the star was massive enough, about six times the mass of our sun, the explosion, while throwing some material out into space, will cause the core to implode into an object just a few miles in diameter, but so massive that its gravity prevents light from escaping. Gravity pulls everything into it as if it were a hole; light cannot escape from it, so it is black—a black hole."

"I've heard that some astronauts entering earth's atmosphere saw flashes of light that were generated by cosmic rays energizing the optical nerves," I said.

"Yes, they saw flashes with their eyes closed," Goswami replied. "The abundance counts of the flashes correlated very strongly with the results of cosmic-ray detectors."

"Sometimes, if I defocus my eyes like this, I see a field of fleeting scintillations." I looked, slightly concentrated, at the tip of my nose and focused on the sparkles of light.

"Yes, so do I," replied Goswami. "And so did the Rishis. They called it the starry sky. But those are not cosmic rays; they are most likely virtual particle transitions."

" 'Virtual particle transitions'? What are they?" I asked.

"They are particles that jump out of the quantum vacuum and fall back," he explained.

"It's like the zero-point energy," I replied.

"Yes, the zero-point energy is the energy of the quantum vacuum. The quantum vacuum, or quantum field, is a field of potential that we can't detect. It's not really a vacuum. In fact, immense energy and information is there, but we can't use it. Quantum field theory regards all particles as excitations of the quantum vacuum."

"Why can't we detect it?" I asked.

"Because we are like fish in water. The zero-point energy is all around us, uniformly distributed," he said.

"Even in ocean water there are some differences, variations of temperature and salt content, even though it looks the same," I said.

"So far we have not detected any variations in the quantum vacuum," replied Goswami.

"What about at the atomic level? Even though water is uniform, we can detect individual atoms that are different—hydrogen, oxygen, sodium, chlorine, and potassium."

"Correct, absolutely correct, but in order to see variations in the quantum field, you have to look at it in a scale a trillion times smaller than an atom."

"You mean there is a sea of energy all around us, through us, pervading everything, but balanced out, canceled so that we are adjusted to its pressure like a fish is to water?"

"That's right. A fish is the last to discover water," replied Goswami.

"If we can't detect the zero-point energy, how do we know it exists?"

"We infer its presence in two ways. First, if matter is continuously manifesting out of background energy, that energy must be immense. Secondly, space is crisscrossed with electromagnetic waves, billions of them from each of billions of stars. We think of these waves as having no energy in empty space, at zero point. Quantum theory predicts a small residual energy in some waves. Add up the amount of zero-point energy in the waves and it is immense."

"And matter continuously forms from this energy?" I asked.

"Yes, and some particles get enough momentum in space-time to temporarily jump out of the quantum vacuum. It's like heating water. Before the water actually boils, molecules are jumping out all the time, but fall back."

"What's that have to do with flashes of light?"

"When the virtual transitions jump out of the quantum vacuum, they give off a flash of light as they fall back. It's happening everywhere all the time," replied Goswami.

"The quantum vacuum seems a lot like the Void our yoga teacher described to us." I fell quiet trying to take in the enormity of what I was hearing.

"Yes, the quantum vacuum and the Void are similar. In fact, the vacuum is filled with every potentiality in the universe. The vacuum replaces the material, jellylike ether of nineteenth-century physicists. The universe is not filled with the vacuum; rather it is written on it and emerges out of it. The material universe is no more than an insubstantial cloud compared to the underlying quantum vacuum."

Two monkeys pranced along the windowsill, rattled the glass, and disappeared. I stood up, walked to the window, and watched them bound across the PRL grounds to a construction site. Women in long, flowing tribal dresses, with white ivory bracelets on their slender brown arms, carried bricks on their heads up seven stories of rickety bamboo scaffolding. They formed a long, continuous line that snaked up the scaffolding, starting from a large pile of bricks at the base of the building. In the trees around the brick pile, a dozen babies hung in cloth hammocks. If a baby woke up crying and the mother was up on the scaffolding, another woman rocked the baby until the mother could make it down and nurse the baby. My mind drifted back to Goswami, who had sat down at his desk.

"Perhaps you will be able to find the smallest fundamental particle in the cosmic rays," I said absently.

"Actually, I am doubtful such a thing exists," replied Goswami.

"What do you mean?" I asked.

"We keep finding smaller and smaller particles. If a particle is too small, the gravity forces becomes infinitely strong."

"Yes, of course," I replied.

"It is more likely that elementary matter is transient, keeps manifesting itself from the background energy and then disappearing."

"Like virtual particles?" I asked.

"Exactly," replied Goswami. "Quantum mechanics suggests there are no elementary particles, but that all matter exists as vortices of energy fields that stretch to infinity, and only when we observe them do they collapse to form particles."

"God, what a lot of work," I said.

"No, not at all; it's spontaneous. The work comes in keeping it from collapsing." Goswami smiled.

I had speculated since graduate school that the changes of reality in the near-death state might be related to the collapse of the wave function. Perhaps the mind does not support the collapse in the near-death state. Then I remembered the Bohr-Einstein debate and the problem of faster-than-light collapse of the wave function.

"The collapse of the wave function must occur at speeds faster than the speed of light," I said.

"Right, the collapse of the wave function is a nonlocal event, meaning it occurs instantaneously everywhere at superluminary speeds," replied Goswami.

"Hence, quantum waves aren't real because nothing can travel faster than light," I said.

"Matter can't be accelerated to the speed of light. Quantum waves aren't necessarily matter; they might be nothing more than information. Recent experiments have shown that the universe is nonlocal, meaning it is interconnected at some level. John Bell devised a mathematical proof, based on a thought experiment by Albert Einstein and his colleagues, demonstrating that the universe is nonlocal. Change one part, and the rest of the universe instantaneously changes, faster than the speed of light, as if it were a single object. Alain Aspect at the University of Paris has actually measured nonlocal correlations in the polarization of light."

"How can the universe be interconnected?" I asked.

"Through phase entanglement. Erwin Schrödinger, the father of wave mechanics, called it the chief feature of quantum mechanics. Some people refer to it as phaselock."

"And what is it?" I wondered if phaselock could explain the interconnectedness I had experienced in Tierra del Fuego.

"When two particles come into proximity, some aspects of their wave functions merge, come into resonance like clocks on a wall with their pendulums swinging in rhythm, and remain correlated even after the waves are separated. Action on one particle seems to have an instantaneous effect on the wave function of a distant particle. Correlated but not connected, no detectable link exists," Goswami said.

"How many atoms in the universe have come close enough to become phase correlated?"

"In a thimbleful of ordinary matter, there are billions of billions of atoms, each colliding millions of times a second. Indeed, during the big bang all elementary particles were probably close enough together to become phaselocked. But only certain types of particles, those that can occupy the same space at the same time, phaselock."

"But this phaselock takes place on the quantum level. It doesn't concern the real world." I was concerned that quantum effects did not influence the macro-world of everyday life.

"Phaselock does affect the real world," replied Goswami. "The recent invention of the laser proved that. In the laser, individual photons are reflected back and forth between mirrors so that they collide and become phaselocked, or pumped, to use the vernacular. In essence, all the individual photons combine into a single wave form."

"Is it just true of photons?" I asked.

"No, electrons phaselock in the state of superconductivity. The electrons act together, don't collide with each other, don't impede the smooth flow of electricity through the lattice of a metal."

"I see."

"Some physicists believe phaselock is a basic property of biological systems. Living systems, including the brain, are characterized by their ability to coordinate complicated processes over the dimensions of a cell or organ. The question is: What is the source of this coordination? Is consciousness, for example, a phenomenon involving a single quantum state across whole regions of the brain? The physicist Herbert Fröhlich proposed that metabolic energy 'pumps' living systems into states of self-organization over macroscopic distances. Molecules in cell membranes then vibrate in phase, in the same quantum state, analogous to the laser or a superconducting solid." Goswami leaned back in his chair and looked reflectively at the ceiling.

"Fröhlich systems are a kind of self-organization?" I had decided to pick Goswami's brain.

"Correct," affirmed Goswami. "Self-organization is common in biological systems. Certain biological forms can only be explained by a preexisting template that the organism or part of the organism grows into. For example, a sea urchin embryo cut in half develops into a full sea urchin, albeit smaller than normal. A piece cut off of a flatworm can regrow the entire animal, and it doesn't matter if the head or tail is cut off; the whole worm regenerates."

"What about DNA? Perhaps it controls the shapes of biological systems," I inquired.

"A simple analysis shows this cannot be so," responded Goswami. "How can DNA control when an oak branches? Slime mold changes its shape over its lifetime, starting out as a sluglike moving creature crawling on the earth and ending up as a flowering stalk, fixed in one position. How does DNA tell a slime mold to change its shape? Even in a single organism, different cells have different shapes. When a lizard's tail is amputated, a scab of blood cells forms and the new tail grows out from those uniform cells. How can the DNA in the scab cells of an amputated lizard's tail control when and where bone cells, muscle cells, skin cells, and nerve cells of the new tail manifest? DNA contains a lot of information, but how can it project that information over space and time?"

"I've heard that self-organization can occur in nonliving systems," I remarked.

"Certain patterns of self-organization are pervasive in the universe at all scales," added Goswami. "The most ubiquitous are waves seen in clouds, the ocean, beach sand as dunes and ripples, all the way down to electromagnetic radiation. The spiral is also common in galaxies, the great red spot on Jupiter, hurricanes on Earth, the Coriolis whirlpools of water flushing down a drain. In all cases of self-organization, matter flows in response to the action of energy. The ratio of the energy of flow to the resistance to flow results in the same shapes at all scales of the universe."

"So the wave functions are a source of information that can guide self-organizing systems," I suggested.

"Exactly," replied Goswami enthusiastically. "And the wave function describes where matter will erupt out of the quantum vacuum."

"So the collapse of the wave function is really a change of information," I conjectured. "Information is mass-less, and therefore capable of traveling faster than the speed of light."

"Not exactly," corrected Goswami. "I like the analogy of a waterbed: you press down one part and another part instantaneously moves, because it is all one waterbed, all one universe connected through wave phaselock. We can't see the connections because they are too small or in a hidden dimension."

"In another dimension?" I queried.

"Yes. It's like this: Imagine a fish in a tank with two video cameras, one from the front and one from the side. One camera sees the head move; another sees the sides move. Each image is a two-dimensional representation of part of the fish. A comparison of the two images suggests that the movements are related in time, as if there is instantaneous communication between the two images. But there is no exchange of energy, matter, or information between the two events, because they're part of the same fish, the same movement. In the same way, there is one interconnected field of information, the quantum potential, that underlies all reality."

"That's impossible to prove," I challenged.

"It's equally impossible to prove that it isn't so." Goswami laughed, winked confidentially, and gave a head nod.

"So our thoughts are a kind of Midas's touch; whatever we think about turns into matter." I stood up, stretched, and walked to the window.

"I think so, but it doesn't matter." Goswami shrugged his shoulders and laughed.

The neurons in my brain were starting to make some new connections. Perhaps during the near-death experiences, I had come in touch with the field of information in the quantum potential and could somehow change the information. In fact, that was the only reasonable explanation—all parts of the universe must be interconnected. Either that or my experiences were unreal. The interconnectedness of the universe circumvented the restriction that matter cannot move faster than the speed of light. It validated Einstein's intuitive insight that there is a deep reality beyond mere chance and that God does not play dice. But what is the connection between mind and the universe?

Suddenly the women on the scaffolding dropped their baskets of bricks and stared down. Men ran about with shovels and sticks.

"What's happening?" I asked Goswami. He came to a window, stood beside me, and looked out.

Not thirty feet from the door to our laboratory, a group of men poked sticks into a hole and banged their shovels on the ground.

"They've discovered a cobra den. Looks like they'll smoke the snake out," said Goswami.

"I thought cobras are sacred. Lord Shiva incarnate or some such thing." Men with torches came running up and shoved them down the hole.

"They are sacred," responded Goswami.

"Then why do they want to kill it?" I asked.

"Because if it bites, you will die." He laughed. The women scurried down the scaffolding, and gathered around the men.

"So it is sacred, yet a menace? Both things? How can it be both things at once?"

"Most things are both things at once; it depends how you look at it. Like Schrödinger's cat," he said.

"Schrödinger's cat?" I was familiar with Schrödinger's work on the wave equation using probability statistics for quantum effects.

"Yes, Shrödinger's cat." Goswami stared out the window at the crowd around the cobra's den. "It was a thought experiment to demonstrate the role of the observer in quantum physics."

"I thought Schrödinger worked with probabilities."

"He did, but that led to the thought experiment about the cat. Suppose there is a cat in a box with a canister that will release poison gas when triggered by high-energy electrons emitted from a radioactive source. The question is this: Do the particles exist only as waves incapable of triggering release of the gas until an observer opens the box and they change to particles? Will the cat be dead or alive when the box is opened? Can the cat exist in an intermediate state half-dead and half-alive? According to the uncertainties in quantum mechanics, you cannot prove whether it is dead or alive, or, for that matter, even in the box until you open it and look at it. It is the act of observation that decides the fate of the cat."

"And you say this applies to cobras. How?" I asked.

"When the cobra is in the den, you don't know if it is Lord Shiva about to give you the blessing of enlightenment or a deadly menace that will attack and kill you when you get near it." Goswami shrugged.

"It depends on how you perceive it," I suggested.

"Exactly." Goswami turned from the window. "And that depends on the situation, your frame of mind. In a sacred cave in the Himalayas, a cobra will definitely be Lord Shiva if you believe it is, but here at a scientific research laboratory, the snake will definitely attack, bite, and kill you." Goswami looked at me, eyes bright with amusement.

"Why's that?"

"Because that is what scientists do." He laughed. "Collapse the wave function. We don't believe anything we cannot measure. We believe the results are independent of the attitude of the observer, but they are not."

"What do you mean?'

"Because of the wave-particle duality. We understand that light is both a wave and a particle, that each description excludes the other. But both descriptions are necessary; they complement each other," asserted Goswami.

"Oh?"

"The wave nature of reality collapses as soon as we try to analyze it, try to separate ourselves from it. No, if you see a cobra here at PRL, kill it before it kills you."

"We can do experiments that indirectly show particles exists as waves, but when we try to measure them directly, we change them," I offered.

"Exactly. For example, if you shine a light on a particle, you change its nature, velocity, frequency, et cetera, so you as the observer have changed the observed," added Goswami.

"So, we can't measure particles without changing them," I responded.

"It is even more complex than that. Complementarity is a fundamental feature of the universe. Position and momentum actually exist combined as a single property; they are complementary attributes. We can only choose to measure one property, by excluding the other."

"I've heard of that about the complementarity of waves and particles, as well as of position and momentum, but there are other types of complementary systems. Right?"

"Yes, time and energy, for example." Goswami headed toward the office door.

"What do you mean?" I turned to him.

"The energy-time uncertainty requires that the intervals of energy measurement should not be too close together or there will be an indeterminacy in the quantity of energy involved. Any energy is theoretically possible for minute amounts of time. For example, if electrons are shot at an energy barrier, some of them can borrow energy from the zero-point energy for a very short period of time, something like a millionth of a billionth of a second. That electron can then surmount the barrier and thus appear to have tunneled through. This is the Josephson effect that can be applied to electrical conductivity over micro distances."

Goswami opened the door and stepped into the hall. "Time for lunch."

"Yes, time for a break." This was all getting to be too much. I followed Goswami down the hall.

"Of course, the converse is true. In the presence of immense energies, time slows down. In the case of black holes such as Cygnus X-2—remembering that mass converts to energy by Einstein's equations—time comes to a halt."

We stepped out into the intense Indian heat, and as we walked toward the dining room, a cheer went up from the crowd around the cobra den. One man spun out of the crowd with the limp body of the cobra dangling from a stick.

"Looks like Schrödinger's cobra is dead," I said.

"As I predicted," laughed Goswami

The man with the cobra on a stick held it out in front of my face. The brilliant sun shown with myriad reflection of the sleek, limp body. What a waste, this vital energy gone, the fierce uncertainty of the snake vanished.

We met Soma outside the cafeteria on the first floor of the main building. Workers and scientists alike lined up to wash their hands and scrub their teeth with their fingers at a row of outside sinks.

"What's for lunch, something hot?" I picked up my tray and slid it along a counter beside dishes piled high with rice and chapatis.

"Oh, no, everything here is very mild." Soma smiled at me and cocked his head. "Try that." Soma gestured for an attendant to spoon cauliflower sautéed in cilantro and cumin sauce onto my plate. I dabbed my finger in the sauce and touched it to my tongue.

"Not hot! Jesus, that burns." I took a drink of water. "It even looks hot!"

Soma scraped a finger full of sauce onto his tongue and let it sit there. "It's all a matter of taste. See, not hot." He picked up a chili pepper and downed it. "Now, that is pleasantly spicy. I can eat three of them in a row."

"I can eat four of them in a row," countered Goswami. We picked up our trays and walked through the cafeteria.

"No need to prove anything to me," I said. We sat down at a table with a half dozen scientists clad in white shirts. Soma and Goswami got into their usual lunchtime bragging competition.

"I am having two punctures today." The bragging started with Soma. He scraped a mound of yellow rice onto a chapati with his hand and bit into it.

"I am having three punctures today." Goswami mixed rice and cauliflower together with a chapati.

"I had six punctures today," I joined in. All the men at the table looked up and stared.

Goswami shook his head and laughed. "That means one every two hundred yards."

"Correct," I replied. "It took me three hours to get to work." Everybody laughed.

One of Soma's assistants rushed up to Soma, whispered in his ear, and handed him a telegraph. Soma looked at it and handed it to me. "For you, from the States."

I opened it and it read, "Regret to inform you, Mom died today. Dad."

I paused, looked around, got up, and stammered, "Excuse me, bad news."

They shook their heads quietly in reply.

I bicycled home with a monsoon of tears welling around my brain. Halfway home the air went out of my tire, and I pushed the bike limply through the stark cactus rows in the intolerable heat. The glare of the sun on the sand gave me a headache that built up pressure in my sinus cavities. I threw the bicycle on the ground outside the front door.

"Why are you home so early?" Maggi met me at the door. I showed her the telegram, went to the bedroom, lay down, and stared at the ceiling. Maggi came in and lay down next to me. God, the cracks and water stains on the ceiling; it was getting ready to cave in. Nothing lasts.

I couldn't stop thinking of my mother. Images of her in housedress and sneakers, picking blueberries on summer days in Lynn Woods, rolled through my mind. I could feel the coolness of the shade as we ate lunch beneath the perched granite boulder we called the cave. We were perfect and happy together then. The orioles and robins sang in the bushes. We knelt under the black oaks, pulling off berries in the summer heat. I loved her and was loved by her. I have been in love with black oaks and orioles and blueberries ever since. We shared perfect vacations camped out at White Lake in New Hampshire. We climbed Mount Chocorua when I was eight. I have climbed mountains ever since. How many times had she helped me, saved me from leeches?

Now, the pressure in my head and throat was immense. I turned to Maggi. She cradled my head, and the tears cut loose with an intensity that could have quenched the thirst of a cactus. I knew we could not make it back to America for the funeral. My mother and I had been far apart at the end. I went to the bathroom and noticed a toad in the bowl. Fuck it, who gives a shit. Without a second thought I flushed the toilet. Maggi helped me upstairs to the roof, and I collapsed on the bedding, sobbing against her. She rubbed my head and stroked my temple.

I sobbed myself to sleep, and when I woke up, Santabehn was sitting beside the bed wiping tears from her eyes. Maggi sat up, talked to her, hugged her, and cried openly.

"What's happening?" I sat up.

"Gopal died." Maggi turned to me, tears streaming down her face.

"Santabehn's baby died?"

Yes," she replied.

"How? He wasn't even a year old." I was flabbergasted. The death of a child is the worst grief of all.

Overhead, the murderous flight of crows began with annoying flapping. Their black forms filled the sky, dark and threatening. They croaked with the reckless indifference to life, accidents of capricious selection in the natural order.

Maggi dabbed her tears. "If Guruji is right," Maggi said, "your mother and Gopal may be better off now, beings of light at the side of God."

"Of course," I said, "but that doesn't make me feel any better. I miss them. Why were they taken away?"

"To help you give up your attachments," she offered.

"Oh, that's bullshit!" I said, "My attachments are me. I can't give them up just like that."

"You're right; I'm sorry, I was just trying to help." She began to sob again.

"I'm sorry too. Thanks for trying to help me." We held each other. Santabehn, wiping her eyes, stood up and left.

"I wish these fucking crows would go away." I stood up and threw a pillow in their direction. There was a brief interruption in their flow. They filled the gap and continued. A bloodred sun blotted with limp gray clouds sank into the desert. Ominous patches of night closed in around us, cataracts on the vision of the supreme mind. We were alone facing eternal night with no hope of a future, no explanation for the insanity and grief of our lives.

Then I felt the baby kick.

CHAPTER TEN

THE PHASELOCK CODE

Herds of livestock wandered the desert around our bungalow in search of the tiniest morsel of grass or sip of water. Goat herders in blue turbans and embroidered vests clicked and hissed instructions as their charges trotted between the cactus hedges, bleating and exploring for anything edible. Cowherds in red turbans draped their arms over long staffs balanced across their shoulders crucifixion-style. White waistcoats danced around their hips like miniskirts as they loped along behind snow-white cows and gray water buffaloes caked with mud. Tribal women in skirts and trim halters scoured the sand for fresh droppings and packed them in baskets carried on their heads. The skin over the rib cages of the livestock sank between the bones as the dry season took hold. Maggi and I offered water from our outside clothes washing stand to whoever came by; it was a way to meet our neighbors.

All the time, Maggi's belly grew.

A tribe of nomads set up camp near our house, their camels braying long and mournfully in the smoke of the campfires. I tried to imagine the distances they had traveled, how they circumvented cities and roads. Two nomad girls with orange veils over their heads stopped by for water. Both wore large nose rings and, when they saw that Maggi was without a nose ring, they pointed at her, laughed uproariously, and clapped their hands. The next day the nomads disappeared without a trace.

Two sadhus appeared from between the cactus hedgerows and led a white cow boldly up to our water faucet. I admired sadhus, holy men who had renounced worldly attachments to wander the lanes and holy places of India in search of enlightenment. I had heard stories of their powers.

"Pani?" The word for *water* was one of the few words of Gujarati I had learned.

The taller of the two sadhus, in orange dhoti and vest, looked at me fiercely with eyes bloodshot from ganja, the Indian version of marijuana.

"Pani? No!" he said sternly, pointing to the hair piled on his head in a beehive like Shiva the Destroyer's. As he shook his head, the black hair unfurled down his side. He grabbed a thick lock in two hands and twisted. Water wrung out, splattered on the ground, kicked up dust, and turned the sand wet.

How did he do that? I thought to myself. Has he been walking around with a bag of water in his hair?

"Panj rupee!" He jabbed his wooden begging bowl at me. All right, yeah, yeah. I fished a five-rupee note out of my pocket.

As soon as the other sadhu, shorter with gray hair tucked under a red turban, saw me drop the note in the bowl, he tapped the cow with his staff. The cow lifted its front feet one after the other in a circular motion. It looked at me, eyes askance, as if to say, "How many cows do you know that can do this?"

"Panj rupee!" The gray-haired sadhu thrust his palm in front of my nose.

"No more." I signaled, passing my hands over each other like a baseball umpire signaling safe at home plate. The cow kept lifting its feet with a silly grin. I kept signaling safe at home. Both sadhus gestured and shouted.

I turned my pockets inside out. "No more rupee, finished, kaput." I gestured for them to leave.

The second sadhu banged his stick on the ground.

"Maggi!" I yelled, anxious and confused.

She came out onto the porch, very pregnant. The sadhus stopped their yelling and stared at her. The cow kept lifting its feet.

"Maggi, I need five rupees to make this cow stop," I said in exasperation.

Maggi disappeared into the house and came out with a five-rupee note and dropped it in the bowl. The gray-haired sadhu tapped the cow with his stick, and it stopped lifting its feet.

The black-haired sadhu, his long hair wild as Rasputin's, picked up a pinch of sand, and as he rubbed it on his palm, it turned from brown to black.

He must have charcoal up his sleeve, I thought, or somewhere. Actually, he didn't have sleeves.

He jumped up onto the porch and pointed to Maggi's belly. Taking a dagger out of his belt, he tilted his head back, opened his mouth like a fish underwater, and swallowed the dagger to its hilt.

Big deal, I thought to myself; any circus clown can do that.

Then he jammed the sword farther, and the tip came out through the black scraggy beard at the base of his throat.

Oh, gross, I thought. I heard Maggi gasp.

He turned slowly, pulling the dagger out. After he completed the circle, there was a small red bruise at the base of his throat, but it was healed, no fresh blood. Maggi watched with her hand over her mouth.

Then he held out his hand and yelled, "One hundred rupee, memsahib."

I was impressed, and not a little frightened. "God, give it to him, Maggi, before he curses us or something."

"No more rupee; rupees finished," said Maggi

"You, baba. One hundred rupees!" he shouted fiercely, and pointed at Maggi's belly. I imagined he was threatening us.

"No, please, we don't have any more money." Maggi crossed her hands over her womb protectively.

The sadhu banged his stick, gestured madly, and stared fiercely, his eyeballs as large as mangoes. "One hundred rupees!"

Maggi held her face in her hand and sobbed, "No more rupees, no more rupees." She looked up at the sadhu, eyes rimmed red and tears on her cheeks. We were honestly out of money.

The sadhu suddenly put down his stick, coiled his hair on his head, and tilted his head back proudly. "No more rupee?"

"Yes. No more rupee," Maggi said quietly with dignity. "Me baba, please."

"No more rupee? *Acha!* You, baba." He pointed to Maggi and pulled a fist-sized stone lingam out of his mouth, handed it to Maggi, and turned on the heel of his hemp sandal.

The gray-haired sadhu tugged on the cow's bridle and, as he

started off, pointed to a smooth, round, potato-shaped stone on the sand. I didn't remember any stone there, and picked it up. The word *POTATO* had been traced on it in water.

"Was this stone here before?" I held it up for Maggi to see. Stones, especially smooth ones, were rare in the desert sand.

"No, I don't remember it."

"Well, look at this." I pointed to the lettering. "What do you make of it?"

"It spells *potato*," she replied.

"I know that, but how can it be?" I scanned the cactus hedgerows, shielding my eyes from the bright sky with my hand. The sadhus and their cow had disappeared as if they had never been there.

"God, Maggi, he looked like he was going to curse the baby," I said.

"No, he was bargaining with us; when he realized we really didn't have any rupees, he gave us the lingam for free." She held out the lingam for me to see. "It's good luck for the child."

"Did that dagger come out the base of his neck?" I rubbed the skin around my neck.

"It looked that way to me," she replied.

"Unbelievable," I said.

"Well, look. Here's the lingam, and you still have the potato rock." She shrugged.

I looked at the rock. The word *POTATO* vaporized and disappeared before my eyes.

The meeting changed my view of sadhus. Their magic was astonishing, but their humanity lacking. The poverty, or perhaps the ganja, made them harsh. I could never hack it as a sadhu. I'd spend all my time worrying about my next meal and bed rest, rather than contemplating the mind of Brahma, the wonder of Shiva, or the kingdom of God.

As Maggi turned to enter the house, her face contorted in pain, she grabbed her back and supported herself against the doorjamb.

"What's happening? Are you . . . ?"

She took a deep breath and slowly straightened up. "Just a Braxton-Hicks."

The monsoon was overdue. During the day, the temperature reached one hundred twenty degrees Fahrenheit in the sun. The animals, expecting change any day, were agitated. The water buffaloes lowed in mellow anticipation of abundant mud wallows; the camels brayed and showed their teeth defiantly to the accumulating thunderheads and heat lightning sparking on the horizon.

Maggi waddled around as huge as a water buffalo and tended to last-minute preparations.

I had developed a serious problem with sinus headaches. During the day, the cavities of my skull filled with fluid; at night a terrible pressure built up around my brain as if the heat of the day were trapped in pockets and throbbed to be released. At first I inhaled the fumes from Vicks VapoRub or Tiger Balm to break up the congestion and relieve the pain. But as the heat of the dry season increased, no medication, aryuvedic, homeopathic, naturopathic, or allopathic, relieved the pressure.

It was May 28, 1974, and Maggi was asleep in the bedroom. She had boiled water, and I was breathing in the usual vapors with a towel draped over my head. On this night the vapors didn't help, and as the cold night air invaded our house, I came down with the worst headache of my life. I sprawled out on the front porch. The heat of the day that radiated from the concrete slab relieved the pressure in my sinuses. Maggi woke up and cried out. I stopped moaning long enough to check my watch. It was 12:20 A.M.

Guruji had told me chanting would relax me. Perhaps it would help my headache.

I tried chanting. *"Aummm. Aummmm. Aummmm.* Oh, shit." The pain was too much. "Oh shit! Oh shit! Oh shit!" I screamed into the night, not caring if I woke the neighbors. The fever brewed in my skull, and the sweat poured off my face.

Was my headache the result of the anxiety that had increased each day as the birth approached? There were the normal worries associated with pregnancy. Would there be complications? Was the child affected by hepatitis that had infected me and, to a lesser extent, Maggi seven months earlier? But my anxiety seemed to come from unconscious levels beyond the normal worries over the fate of the child or health of the mother.

Perhaps I wasn't ready to be a father. How many people had told me, "Having a family, that will change the old lifestyle!" Perhaps I couldn't cope with giving up life on the road. The reality impressed on me from my first breath had been that material success was the essence of being a good parent. That was my conditioning, and it rose, unsummoned, to joust with my intellect and reasoning. I would have to get a steady job. Although I had managed some research in the post of minor professorships here and there, I hadn't settled down in one position. I was interested in carrying out research into the distortions of reality I experienced during my near-death experiences. No university would fund research into near-death experiences. I would be labeled as unreliable, erratic, and unstable. So I carried out my research independently.

All kinds of thoughts raced through my mind. In my grandfather's day, happiness and validation came from the craft of a job well done. Can a child find happiness or joy or even survive in the consumer culture after watching twenty-four thousand TV ads? Purchasing power was no substitute for unconditional validation from a father. How many fathers impose their fantasies of material security on their kids, expecting their children to validate their own way of life? I didn't want to do that to anybody.

Maggi's instincts were more positive. She wanted kids. Her soul pushed toward that inevitable goal without hesitation. Each child is sacred, according to Maggi. I counted on her to guide me into parenthood.

Right now I needed to throw up in the worst way, but hating the taste of vomit, I held back. I heard barks and saw a pack of jackals converge around a nearby streetlight, yelping and snapping at the swarming insects attracted to the light.

My stomach twisted in knots, and finally, with a loud scream I threw up. The jackals scattered into the shadows. The tension left my body with my stomach contents. I rolled over on my back and wiped my mouth on my sleeve. The knot of muscles around my skull unraveled.

"Oh, God, that's better," I murmured, and looked up at the stars. The familiar constellations blazed brilliantly. The eagle and swan flew straight overhead. I thought the stars in India would be different, more exotic, but they were the same, with different names.

Maggi groaned from the bedroom.

"Maggi? Are you all right?" I stood up and ran into the house checking my watch.

Maggi lay on the bed, her hand on her belly. She looked up at me, arched her back, and let out a loud groan.

"Oh, God. Oh, God, Oh, God," she cried. Then she crumpled back on the bed and relaxed.

"Thirty seconds. I don't think that was a Braxton-Hicks." For nine months the fetal placenta had been producing a hormone that kept the uterine muscle relaxed. Evidently, production of the hormone had just stopped.

"Neither do I," she replied.

"I'll go for the car." I kissed her hand and stood up. A soft quarter moon had passed the zenith and was sinking in the west. I pedaled my bicycle with a fury through the cactus hedgerows, following the faint cone of jiggling light from the handlebar. No time for a puncture now. I skirted around potholes, climbed over hummocks on the dirt track, and fought to keep the bike from swerving.

I swung by Santabehn's hut and knocked on the wooden door. When she sleepily poked her head out the door, I said in pigeon-Gujurati, "Memsahib, baba." I pointed to the house.

"Acha." She nodded, closed the door, and retreated to the darkness of the house.

Arriving at the laboratory, I pedaled up to the chauffeur's station and signaled with my hands that my wife was having a baby. The driver jumped into the car, and I jumped into the passenger seat. We pulled up in front of the house and I ran in. Santabehn held Maggi, squatting in the tribal lady's birth position, ready to have her baby right there and then.

I helped her up and to the car, and we sped through the dark, narrow streets of Ahmedabad. As we passed the low adobe huts of tribals, I thought, how many children sleep inside? We passed row after row of tenant buildings with the plumbing crawling over the walls, mildew and mold hanging down near leaks.

Maggi screamed, arched her back, panted, and grabbed me by the shirt.

"Forty-two seconds. Remember to breathe."

"You try breathing. It doesn't help." She collapsed back onto the seat of the car.

"Remember what the Lamaze instructions said. Pain is all in the mind. Don't expect it, and it won't come."

"Just like during your headaches?"

"Good point. I know, easier said than done. I was just trying to help." I rubbed her back.

"The distraction does help a bit." She reached out and took my hand.

In fact, the contractions that followed didn't seem as distressing. I speculated that now that she knew what to expect and focused on her breathing, she noticed the pain less.

We passed thousands of tenant families, maybe five thousand sleeping children. Where did these families summon the courage to rear children in utter poverty and lack of security?

We pulled up to the birthing house, a three-story building with balconies and arched windows. Nurses in white saris helped Maggi to a comfortable, soft-lit room. Children, women, and men, whole families, slept on mats in the halls like in a railroad waiting room, baskets of food around them. Birth in India was a family affair; everybody slept over. Maggi lay in bed, and soon the doctor's aide came in and checked her dilation. I held her hand, and stroked her forehead.

"Oh, God, why did I ever think I wanted to have a baby?" she said.

"It's okay. It's going to be okay."

"It's just two centimeters. It could be some time," the nurse said. "Make yourself comfortable, and we'll wait."

I climbed into bed beside Maggi. She stirred every ten minutes or so, and I held her and rubbed her back. At dawn she got up to go to the bathroom.

She came back to bed and announced, "My waters broke in there; it won't be long."

A white-haired nurse came in and checked the dilation. "Five centimeters." She smiled and patted Maggi's hand. "Try to get some rest. It'll be a while yet."

The next contraction was fifty-two seconds long, four minutes before the next.

"Time to go out for breakfast; anything special you would like? Chapatis? Doughnuts? Juice?" I asked.

"I am thirsty—I know—ice cream!"

"What if you need anesthesia?"

"I know I won't."

"In that case absolutely, just the ticket. Now, don't have this baby until I get back." I knew an amazingly well orchestrated chain of events had to occur. The bones of her pelvic girdle needed time to open, the muscles time to stretch.

As I walked back into the room with a cone of chocolate ripple, Maggi let out a scream of pain. Her face puffed up. She was perspiring heavily.

"It's okay." I mopped her face with a damp cloth, and held the ice cream cone to her mouth once the contraction subsided.

"But what if I can't do it?" She licked at the ice cream; her forehead was creased with doubt.

"You can do it. You're doing fine. Keep it up."

"But it's going to get worse. I don't know if I can stand it."

"Let go. Focus on what's happening now. Don't worry about what's coming. It's okay; it's best if you relax." She took the ice cream from me.

"You know, most places wouldn't let me eat this."

"But you're not going under anesthesia, remember?"

"I'm not sure anymore; I'm not sure I can stand this pain."

"Well, how about a little painkiller, no?"

"Perhaps."

"I'll ask the nurse." I went to the nurse's station. "Do you have anything for pain?"

"Yes, Demerol, if she needs it."

"She won't be too dopey?"

"Well, a little, but she'll be fully conscious and able to push."

"Sounds good."

I returned to the room just as another contraction peaked.

"Oh, God, I can't stand it." Her legs quivered with pain. She clutched my forearm. Her fingernails turned white and pressed the flesh on my arm to the bone.

The nurse came in and gave her a shot, examined her, and announced brightly, "The baby has dropped; everything is fine."

At noon the contractions were about sixty seconds and almost on top of each other. The doctor made his first appearance, checked the dilation, and announced, "Ten centimeters; time to go."

Three nurses helped Maggi into a smock and onto a wheeled stretcher, and they pushed her into the birthing room.

A nurse helped me into a surgical mask and hospital smock and led me to the birthing room.

"The baby has turned," said the nurse, pulling back Maggi's smock.

"It won't be long." The doctor probed her cervix.

The contractions were almost a minute long with several minutes between. They did not seem as painful. On the next contraction I could see a pale spot of flesh crowning, and during the following contraction, the spot increased from the size of a nickel to the size of a half dollar.

"We are ready to push," said the doctor.

Maggie was pushing and groaning, and the arteries on her forehead and neck swelled and throbbed. I felt totally useless. Even my attempts to comfort her were to no avail. She let out a loud grunt that hung in the air for an eternity and would have curdled a grizzly bear's blood. It wasn't a scream of pain but more like the noise of hard work being done.

"The head is crowning," said the doctor.

"Why did I let you talk me into this?" she gasped.

"You're doing fine." I mopped her neck and forehead with the damp cloth.

Maggi bellowed as if doing a 330-pound dead lift and focused every fiber of her being on the push. Her cheeks puffed up, and her face turned red as a beet. I grabbed her hand, and she pierced my palm with her fingernails.

"The baby is right here." I pointed to the baby's wrinkled head.

Maggi reached down and felt the tiny head with her fingers. A nurse held up a mirror so Maggi could see. She bellowed one more time, and Brendan slid out wet, silent, and squirming as if swimming underwater. It was 2:02 P.M. He was shining in a clear bright way, all covered with gleaming white vernix. Maggi collapsed back and cried softly. It was amazing, astounding, and incredible, all due to Maggi and Brendan. The doctor, nurses, and I were no help whatsoever. His will to be alive, his determination, was astounding; her courage, immense.

"It's a boy." I wiped Maggi's forehead, knelt down, and hugged her. "You were wonderful; thank you for working so hard."

She looked at me warily, too tired to answer, her eyes puffed up as though she'd been through a prizefight.

Time stood still as we waited for his first breath. The doctor cradled Brendan in a swaddling cloth. A calm potential filled the room with a sense of lightness. I felt peaceful, much as I had been during my near-death experiences. The doctor cleared the mucus from his mouth and nostrils and patted him dry with a towel. Expressions twitched across the muscles of his face like shadows chasing each other on a cloudy day. I knew unseen miracles were happening. Brendan's umbilical arteries were automatically shutting; the umbilical vein was waiting just long enough for the blood in the placenta to return to him. Blood flow was being rerouted through the vessels of his lungs and intestines. A hole between the two sides of his heart was closing as the direction of the flow of blood through the heart reversed. Arteries and veins were rerouted as his stomach prepared to digest food for the first time.

Then I had a momentary doubt; oh, my God, he's dying. Suddenly his tiny puckered body filled with breath; he arched his back, raised his arms next to his ears as if to block out the sound, and let out a raging howl. Like the sun burning a hole in clouds, the deep red color of his own circulating blood chased away the bluish tint from his chest, head, and extremities.

The doctor gently laid Brendan between Maggi's breasts. He calmed down immediately.

To think that I had endured the process of being born was incomprehensible. Who would want to go through this? Maggi touched Brendan on a cheek, and he moved his head back and forth, rooting for a breast. She directed him to it, and he began to nurse, gently kneading Maggi's breast as if he had been instructed in a nursing support group. I knew that nursing would stimulate natural hormones that would help contract the uterus, deliver the placenta, and prevent bleeding. I marveled that the cord was just long enough to allow for these happy events. It seemed to me that, in celebration and recognition, there should be a standard unit of measurement designated as the *umbilicum,* the distance from womb to breast.

Maggi, eyes closed and dead-tired, gently caressed and cuddled him, even though he was still covered with patches of vernix and blood.

The doctor peered intently under Maggi's smock. The umbilical cord tightened as he gave a couple tugs. "Looks like we have one small problem," he said.

"What's that?" I stood up

"Nothing to worry about." He clamped the cord and handed me a pair of surgical scissors. "You will do the honors?"

"My pleasure." I cut the cord, plagued by the nagging feeling that I was being patronized. When someone tells me not to worry, that's when I begin to worry.

After he had nursed his full, Brendan turned sideways, opened one tiny eye, and peered for the first time at the world. He moved first one arm through space, then the other, as if trying to play patty-cake but unable to judge the distance between his hands. He was measuring, getting a feeling for space.

A nurse lifted him and gave him to me to hold. God, he was so tiny, so precious, so vulnerable, so brave. No adventure I had ever been on compared in bravery to this single act; no creativity, act of art, of science, compared in fullness to the richness of this creation.

He curled and uncurled his fingers as if he had an exercise ball, and looked at me with one eye. The other was glued shut with mucus. I wiped it off, and he gazed at me with two bright eyes wrinkled with pleasure, steady, comprehending. The corners of his mouth seemed to lift in a faint smile.

He scrutinized me intently, then turned his head from side to side, his eyes moving from one object to the other. He gave exclamations of surprise and joy and occasionally kicked his feet. His eyes returned to me. An unusual affection passed between us that has never been broken. I touched his hand with my finger, and he held on to it tightly. The room tingled and buzzed with tiny flashes of light. I filled with more love than I had known my whole life. As much as I loved Maggi, loving this baby surpassed my expectations. I wanted to protect him, to help him cope with the world. My anxieties about being a parent faded in the moment. Maggi was right; children are sacred.

One by one the nurses pulled the blanket off his face and stroked his cheek.

"Oh, Maggi, you did it. He's the most beautiful thing I ever saw." I bent down and kissed her. "I love you." She nodded and smiled weakly, the purple color had left her face. I laid Brendan on her breast, and he resumed nursing. As long as the child was with us, wherever we went, we would be a family together.

Maggi draped her arm over Brendan and instinctively stroked and caressed him. I felt humbled and wonderful at the same time. I was humbled because this marvelous act of creativity belonged to Brendan and Maggi; at the same time, holding Brendan was the greatest feeling of my life. I had been part of the most wonderful thing I had ever witnessed.

Maggi winced with pain, and we both looked toward the doctor, kneeling at the foot of the bed. "The placenta delivered fine, but there's a couple tears to suture up," he said.

When he finished, there was general exhilaration. I felt as high as a kite. We had made it through. The doctor and nurses congratulated us. They were high and teary-eyed. I wondered if this was the result of the ordeal of labor being over. But it wasn't that much of an ordeal for them. Brendan's effort to take on life made us all happy we had been born. If as mortal humans we are destined to experience holy moments, this was one of them. Somehow it seemed Brendan had brought energy and love from another plane of existence.

One thing I was sure of, the timing involved in the birth process, from the molecular to biological levels, was unbelievable. Inducement of events by hormones was not enough of an explanation. How did a particular hormone know when to go into action? There had to have been a metaphysical timing mechanism behind it all.

It became clear to me that my own birth process must have conditioned many of my own worldviews. My anxiety, my striving for success, my quest for recognition, were quixotic attempts to recapture this moment of intimacy with my mother at birth. Many of the differences between Western and Eastern cultures, between developed and undeveloped countries, could have begun with the birth process and been perpetuated by a basic insecurity generated during several centuries of barbaric birthing processes carried out in arrogance by male doctors in Western society.

Several days after Brendan's birth, the first rain of the monsoon landed on the tomb-dry soil surrounding our house. The first drop darkened the earth and kicked up a crown of dust. Another dark spot appeared after the first one was already dry. Two drops fell nearby, and, before they dried, a splattering of drops sprayed the earth like buckshot. Incredible walls of thunderheads lit up as lightning cracked within, thunder shook the earth, and rain fell with a steady patter. The sand darkened, pitted with craters like smallpox scars. At first the water pooled on the surface, but when the gates of heaven opened and poured out a deluge, the water collected in gullies and rivulets that slithered across the sand seeking the lowest point, forming streams that ran to the storm drains and into the Sabarmati River. Schoolchildren scampered home holding their books over their heads. A warm smell of fertile land filled the air, and three days later the desert sprouted a carpet of grass.

Every morning a woman came to our house and massaged Maggi and Brendan with hot mustard oil. She stretched Brendan in various positions with great care and, cupping her hands under his head, swung him back and forth. He made steps with his legs as if walking in the air.

We were lying on our rooftop bed, watching the crows wing over. Brendan was asleep in a wooden-swing cradle, a sawhorse-like frame of polished and ornately painted wood with a cradle of dowels hanging like a swing from the A-frame.

The music of *bhujans* filled the air. A chorus of women from Santabehn's village jubilantly sang, "Hare Krishna, Hare Krishna, Hare Krishna," clapping their hands to the rhythm of a drum.

"Incredible, as hard as those women work, they still have the energy to sing at night," I said.

"Perhaps the singing gives them energy," Maggi replied.

A peacock croaked from a mango tree.

"See"—Maggi smiled—"the peacock is Krishna singing along."

"It's just a coincidence."

"So what if it is; they believe it's Krishna. What's wrong with that? It makes them happy."

"Well it's okay," I said. "I'm glad they have faith, but I can't go along with blind faith."

"What do you mean?"

"I can't accept dogmatic religion," I replied. "Religions start out with an expanded awareness and then collapse into narrow-minded dogma."

"I don't understand," she said.

"For example, I read about Guru Nanak, the founder of the Sikh religion, in a comic book. Some Brahmans were splashing water toward the sun. 'Why are you splashing water toward the sun?' Guru Nanak asked. 'Because that is where God lives,' replied the Brahmans. Guru Nanak started splashing water every which way. 'Why are you splashing water every which way?' asked the Brahmans. 'Because God is everywhere,' replied Guru Nanak."

"And so, what's your point?" asked Maggi.

"Guru Nanak founded the Sikh religion."

"Yeah, and?"

"Remember the night we slept at the Sikh's Golden Temple in Amritsar?" I asked. "That big Sikh guard clubbed our feet because they were pointed toward the holy book."

"I remember that; he looked fierce."

"He considered it improper because our feet pointed toward God. You see, religions start out saying God is everywhere and end up saying he is only in one place."

"Maybe God is everywhere and in one place at the same time."

"That isn't logical," I said.

"Neither is God," she replied.

We watched the sun set, and Venus, the evening star, flared in the sky behind the nightly flight of crows.

Brendan woke, crying up a storm. Maggi picked him up, held his head in the palm of her hand, and tried to nurse him. He cried even louder. "SShh, SShh, it's okay." She bent him over her shoulder, bounced him gently, and when he didn't stop crying, she placed him carefully in the crib and rocked him. He cried louder and flailed with his arms and legs.

"Shush Shuss, what's the matter?" She picked him up and sat on the bed trying to nurse again. He cried more frantically, arched his back, and thrashed about.

"Oh, God! Oh, God! I don't know what he wants." She began to cry.

"Here, give him to me," I said.

"I don't know what to do!" she sobbed. "I don't know what's wrong with him!"

I lifted Brendan from her and walked him around the room cradled in my arm. A few spasms passed through his tiny body and he stopped crying. Walking calmed him down. He's a ritual walker like me. He twiddled his fingers, and looked around at the light over the door and back at me. His eyes, nose, and mouth were so perfectly shaped and seemed to be set peacefully in his smooth face. I held him up to see the crows. He reached toward them, squealed, and kicked his legs.

I carried him to the bedroom and put *Skull and Roses* on the tape deck: Jerry Garcia sang, *"Goin' down the road feeling bad."* When he first heard the music, his body tensed all over with a jerk. He stuck out his arms as if to embrace something. I turned down the volume, and he settled down. I danced and spun around the room as though I were at the Electric Circus. I swung him in time with the music. He laughed and kicked his legs. I blew raspberries on his belly, and he giggled.

Soon he was asleep. I laid him in his crib. As his head hit the mattress, he jerked to consciousness and then fell sweetly asleep again as I rocked the cradle. There was no better vocation in life than rocking a child to sleep. He was teaching me to care.

Maggi lay half asleep, her arms over her eyes. I could see that she had been crying.

"He's asleep," I said.

"I feel like such a failure," she sobbed.

"No, you're not; you're incredible. Giving birth was the hard part; I'll help you take care of the child. Don't worry." I lay down next to her and put my arm over her.

"Thanks, thanks for your help." She pulled my arm in close to her breast and held it.

"What was the birth like for you?" I asked.

"It was amazing, as if my body was taken over by forces beyond me. My muscles went through patterns in concert; they joined in rhythm, all working together. At times the symphony of movement was pleasurable, almost as if I wasn't there. When the pain got really bad, I retreated to a place faraway. My uterus seemed separate from me as it did its work. When the pushing started, I was very much there and that was the hardest work I have ever done. It totally exhausted me."

I looked over at Brendan's crib. Three crows had lighted on the frame and now stared down at him, turning their heads from side to side.

"Oh, God, they're going to peck his eyes out or something!" She jumped up to shoo them away.

"No, wait, Maggi. It's okay," I said.

She froze in position, standing over the crib. Another crow flapped down, landed on her shoulder, and croaked softly in her ear. Small compared to American crows, it had a gray body with black wings. Before we realized what had happened, it had gone, disappeared into the river of crows overhead.

Maggi, carrying Brendan in a side sling, and I climbed out of the motor rickshaw. Scattered sandals lay outside the door at Guruji's ashram. Six men surrounded Guruji on the carpet under his portrait. One after the other, they bowed before him and laid a rupee note at his bare feet. He patted each on the head and said, "Narayan."

The men of the yoga class practicing asanas looked up when they saw Brendan. We knelt in front of Guruji, and I laid a hundred-rupee note at his feet.

He brushed each of us on the head with his hand and said, "Narayan."

"Narayan," we answered.

He handed back the hundred rupees and said, "You are far from home. You need this more than I do."

More than any great act of spirituality, the fact Guruji refused to take money from us endeared him to me.

He rubbed his hand over Brendan's bald head and said, "Narayan, he looks like Eisenhower." The men chortled with laughter.

"My prediction?" Guruji smiled. "It came true?"

The students laughed and clapped their hands, Maggi and I laughed, and Brendan woke up and looked around, bright-eyed.

"What prediction?" I asked.

"My prediction that you would have a son born on May twenty-ninth. It was true!" Guruji laughed.

"Yes, right. Your prediction came true." In my preoccupation with Brendan, I had forgotten Guruji's prediction.

"And?" He looked at me expectantly, as if I was about to be struck down by shattering enlightenment, wrapped in new reality, or at least aroused by a mild epiphany. I felt nothing. As with so many similar experiences, I had discounted the prediction as coincidence, ignored it as irrelevant, and considered it unimportant compared to my everyday problems.

"Okay, I'll bite. How did you do it?" I was consternated.

"You promised you would give me your attention if the prediction came true," Guruji said with great patience.

"Yes, I did," I said. "And I will."

"You promised to give my hypothesis the same consideration you would give any scientific hypothesis, no matter how strange it might seem." The men in the room looked at each other, then at me.

"Come! Come!" Guruji stood up and led us into the meditation room. The men returned to their asanas. The smell of rose incense filled the air in the darkened meditation room; candles burned steadily on an altar in front of pictures of Krishna and Shiva in the dance of the world. We sat down, cross-legged, on the floor. As Maggi laid Brendan across her lap, he gave a start, then fell back asleep.

"May twenty-ninth is significant for you." Guruji sat dignified in the lotus position on the rug opposite us.

"Your out-of-the-body experience on Mount Everest occurred on May twenty-ninth, the precognition of meeting Maggi in the movie theater—May twenty-ninth, the car accident in Morocco—May twenty-ninth, Brendan's birth—May twenty-ninth. Is it a coincidence they all happened on the same day of the year?"

"I don't know . . ."

"Think about it. Aren't there common features to all these experiences?" Guruji looked steadily into my eyes with warm compassion.

"Yes, a sense of calm, a slowing down of time, a sense of detachment," I answered. Maggi and I both stared at him.

"Have you ever had those same feelings on any other day of the year?" Guruji asked.

"Not as strong. Yes, the blizzard in Tierra del Fuego, it wasn't on May twenty-ninth."

"Do you think it is a coincidence that the four out of five times you had these extraordinary feelings was on the same day of the year?'

"No, I suppose it is beyond the possibility of chance."

"So you agree that unusual things happen to you on May twenty-ninth—beyond the possibility of chance?" He shrugged and held his palms up.

"It would be hard to quantify but, yes, beyond the possibility of chance."

"Your horoscope shows May twenty-ninth is an important day for you." Guruji pulled my horoscope out of a wooden box beside the altar.

"An important day?"

"Yes, on that day the sun crosses your rising sign." He pointed to the symbols on the chart.

"You mean my rising sign is a source of energy that affects me?"

"Not exactly; the horoscope tells you about patterns of energy in the Void."

"The energy of the Void?" I had discounted astrology because there was no scientific evidence of forces connecting the planets and stars to humans.

"Patterns of energy in the Void overlap and erupt simultaneously throughout the universe," replied Guruji.

"Simultaneously throughout the universe?" I asked.

"One location is linked with another without crossing space, without delay, without decay. The movement of wall lizards, the flight of a crow, the appearance of a comet, the dawn of a rainbow, all convey information about the patterns of energy in the Void."

During the months in India and through conversations with the physicists at PRL, I had come to equate the Void with the quantum vacuum.

"You mean everything is connected at some level?" I asked.

"Yes," replied Guruji. "Consciousness is composed of vibrations, and all matter to some degree is conscious. We and the stars are part of the same field of vibrations. Separation is only an illusion."

"I can see the possibility that everything is connected through a field of information in the Void, but how does that enable one to see the future?" I asked.

"It's like this." Guruji's eyes were full of delight, humor, and love. "When thought stops, time stops. If you immerse in the eternal present, then there is no difference between this moment and the next."

"You mean the future can be predicted because at some levels time doesn't exist?" It was beginning to make sense to me. If time didn't really exist, then the future would be present now and it might be possible to make contact with it.

Guruji laughed and shook his head in affirmation. "Precisely; there is only one instant of time, and that is *now.* God is continuously asking us if we want to be enlightened this instant. We say, 'No, not right now,' and that thought creates time."

"So there is no absolute time?" I asked.

"Correct! How could God choose a particular moment to create the universe?"

"What do you mean?"

"Time, as we normally think of it, is measured by the movement of one object relative to another, the hands on a clock, the moon around the earth, the earth around the sun. Imagine you are a grain of sand in an hourglass and time is measured by the movement of the individual grains. In the upper body of the hourglass time is imperceptible, most of the grains do not appear to be moving. However, as they approach the neck of the hourglass, individual grains pick up speed and time rushes on. If we look at the universe as a whole, there is no time; it is like the upper body of the hourglass. However, if we focus on a single grain of sand in the universe, then time rushes by."

"Wow, there is no absolute time!" It was one of those views from the top of the mountain that expanded my horizon forever. Once I understood and accepted the idea that there is no absolute time, the meaning of the May 29 events began to fall into place.

"Can anyone experience the change of time?" I asked.

"We all experience the feeling of eternity at one time or another, but we think our cares and sufferings deserve more attention."

"So when we are connected to this timeless place, we can see the future?"

"The future and the past. Certain higher levels of mind have a direct connection across time because they are themselves part of the patterns of energy in the Void."

"Why have I never heard that idea before?" It seemed to me that, in scientific terms, Guruji was suggesting that our minds, or at least part of them, exist in the quantum potential. Quantum mind . . . I liked the idea. It made more sense than the idea that magnetic forces emanate from our brains and connect us to other things.

"We stay in the same reference frame day after day because we made an agreement somewhere along the line—we have even forgotten when. Do not be afraid." He placed the palm of his hand on my forehead.

I felt a surge of heat enter my frontal lobes. I felt incredible love for Maggi and Brendan. I thought of my fall on Everest, the sense that time slowed down and stopped.

I looked up at Guruji. "During my fall on Mount Everest, I seemed to hang in time like it had slowed down or even stopped, but as soon as I became aware that it slowed down, it sped up again. I crashed down into a crevasse."

"Yes, during your fall, you gave up your normal perception of the world. You observed yourself falling down the mountain. You the observer became the observed!"

I followed the warmth of that idea into my brain and closed

my eyes. The starry night engulfed me, and I broke free, floating like during the fall on Mount Everest, surrounded by clouds that moved and shifted like during the blizzard in Tierra del Fuego. I changed direction at the merest thought, rushed toward the ever-changing vanishing point of my perception, and, as in the accident in Morocco, poured through it into a tranquil sea of light. Guruji, Maggi, and Brendan were objects of scintillating waves, brighter than light. Waves and patterns of light connected the four of us and spun off into the universe. We were at the calm hub of time; paths of light radiated out to the past and future. Oh, my God, this is it, I thought. Then the scene vanished from my perception.

The smell of roses consumed me, and I opened my eyes. Guruji had taken his hand off my forehead. Guruji smiled knowingly, beyond words. We sat speechless in the steady light of the candles.

I knew I was supposed to stay in the moment, but I couldn't help myself. "Why can't I stay in that place all the time?" I asked.

A look of consternation and reproach swept Guruji's face. "As soon as you become self-conscious, say 'That's great,' you have empowered the ego and the illusion of time returns."

"I thought enlightenment was permanent," said Maggi, her voice as soft and calm as the darkness.

"No," replied Guruji. "It's a series of small awakenings, two steps forward, then one backward. You have taken a step forward. But you will forget the experience and indulge in your worldly preoccupations."

True, I had shunned my experiences for several years. When I paid attention to them again, they stopped. Shrikes squawked in the trees outside the ashram.

"We are at the same time separate from the cosmos and at one with it. Participants and observers at the same time," Guruji continued. "In everyday life, we experience the separateness. When thought stops, the illusion is broken and we experience the oneness. You Westerners have substituted the illusion for the reality and feed the illusion with a quest for material success that keeps you from the true meaning of life."

Shades of Chombi, I thought. "Nobody's going to believe this," I said, "especially my father."

"You will be accused of being crazy, egocentric, conceited,

heretical, and sinful. But those are judgments from egos that have a vested interest in the illusion we all get lost in. In the end, the truth is the truth. When the ego is set aside, you have access to the totality of memory and time. It is your responsibility to disempower the ego, not only yours but others'."

"It seems difficult. I only get beyond the ego in near-death experiences," I said.

"I know, but the boundaries we erect are conveniences," replied Guruji. "We can unmake them as easily as we made them."

"When I had the precognition of meeting Maggi in the theater, I saw the future, like you predicting Brendan's birth," I said. "What happened to my ego in that case?"

"Your grief disempowered your ego and the illusion it had created, giving you access to a higher dimension beyond space and time."

"A higher dimension?" I asked.

"Yes, it is like this. Imagine you are walking with a friend on a trail that winds through a mountain range. The world around you is a 'normal' three-dimensional world we call reality. The mountains emerge out of the vanishing point and move past you as you walk, but they block the trail ahead from view."

"I like walking through mountains." I looked up at Guruji and smiled.

"I know you do," said Guruji. "Now imagine you are taken aloft in a hot air balloon without your friend knowing it."

"Okay."

"You can look down on your friend walking along the mountain path."

"Right."

"But as you gain elevation, his surroundings become flat to you, his three-dimensional reality reduces to two dimensions from your point of view."

"I see that."

"*Acha,* now you can look along his path and see features that he will encounter in the future, features he can't see."

"Sure."

"In that sense, you can foresee his future because you are observing him from a higher dimension."

"Yes."

"And, when you return to the three-dimensional path, you can predict the future turns in the trail and the objects your friend will encounter. In the same way, you have access to higher dimensions from where you can view the future."

"But doesn't the ability to see the future imply that there's no free will? If the future is predictable, we don't have the freedom to change it," I said.

"The future is substantive enough for us to perceive it, but malleable enough to be susceptible to change. For example, perhaps the trail branches in front of your friend, leading to a cascade of branching trails. You can only give the probability of which trail he will choose, but nonetheless you can predict that he must choose between certain alternatives. The nearer the future, the fewer the alternatives."

"When the future changed in the Tierra del Fuego blizzard, the rain dance at Woodstock, the accident in Morocco, did the future change for everybody else?" I asked.

"All the other trails, all possible futures, still exist whether or not your hiking friend chooses them. God has given us the freedom to choose from an infinity of possible futures. All possible futures coexist. By choosing one, we do not change the others," replied Guruji.

"If there are limitless futures, why do we choose only one?" I asked.

"The limitation of our perceptions and the illusion created by the ego prevent us from being aware of more than one future at a time."

"Why, then, during near-death experiences?" I asked.

"The internal dialogue of the ego stops, you overcome the limitations of your perceptions, a mental screen lifts, and you see beyond the veil of matter to the infinite sea of light. However, the possibility of seeing the future is always with us."

Guruji's argument made sense in terms of the wave-particle duality. The ego exists in the world of matter, the particle phase. On the other hand, the quantum mind experiences the wave functions connected to the quantum potential. But what good was it to know the future if we couldn't experience it all the time?

"Why can't I overcome the ego except in extreme experiences?" I asked.

"The ego doesn't give up its power without a struggle. It follows us like a dark shadow," Guruji replied. "Learning to set the ego aside happens in stages."

"Is there any other way to overcome the illusion created by the ego?" I asked.

"This child will ground you and awaken you to your illusions." Guruji rubbed Brendan's bald head.

"What do you mean?" I asked.

"This child and others to follow are your path to enlightenment. They will teach you to live with your shadow side, to face the opposites, to accept the shadow in others, and to forgive others, especially your own father. The child will show you that most of your perceptions are childhood projections of the fear of abuse and desire for love. Your whole world is conditioned by your desire to be loved by your parents and peers. But in your society, love and validation are commodities, things to be earned or bought. You will become aware of the illusion that is your life, take responsibility, and regain control," said Guruji.

Brendan woke up and started to cry.

"Is it all right if I nurse him?" asked Maggi.

"Absolutely; nursing is natural," replied Guruji.

"The child will teach you that being a parent is the most difficult endeavor on earth," Guruji continued. "The fear of losing one's child grips you like no other force and compels you into all sorts of directions you have sworn over and over again you would not go. Through the humility and suffering you learn by the inevitable mistakes you make, you will learn to forgive and accept the mad schemes, projections, crimes, and wars waged for the children of the world. You will find out that raising a child the way you want is an impossible task."

After a few months, Brendan took to crawling out of the house and visiting the neighbors. But rather than scolding us for not watching him, the neighbors entertained him and painted his eyes with kohl, without reproach or blame. Everybody helped watch after the children in the neighborhood. I felt part of the community, respected and friendly toward everybody. I relaxed as I had never been able to do in America.

One day we were sitting on the roof of our bungalow at sunset. The usual crows flew over.

"So," said Maggi. "Is Guruji right?"

"Right about what?" I asked. Brendan yawned and stretched in his crib.

"Is it possible to tell the future?" asked Maggi.

"Of course not," I replied. "It doesn't make sense."

"Then your May twenty-ninth experiences are just coincidences?"

"That doesn't make sense either."

"Then nothing makes sense?"

"No. I mean, how is it possible to know the future?"

"I don't know. But it seems that it is," she said. A rocket flared up in the sky and blazed over the Physical Research Laboratory.

"There is no scientific proof that it is possible to tell the future." I stood up and scanned the horizon.

"Guruji's prediction of Brendan's birth is not proof enough?"

"Perhaps it was just the power of suggestion. Even if I acknowledged that it is possible to know the future at times, there is no scientific explanation for knowing the future. That's what really counts. I won't believe any of this until I find a scientific explanation." I heard the putter of a motor rickshaw on the road.

"That's your job. You're this great scientist; come up with an explanation. Just because there is no explanation for something doesn't mean one doesn't exist. Isn't that how new discoveries are made? Coming up with explanations for the unexplained?" Fireworks whistled and exploded from all directions.

"Yes, of course. But even if I found an explanation, there is the problem of free will."

"Free will?" She lifted Brendan up and settled him in the sling she carried over her shoulder. He gave a start, then fell asleep against her blouse.

"Yes, if we can know the future, then it must be predetermined and there is no free will."

"What about Guruji's explanation? The branching paths thing?" asked Maggi.

"Well, that works as a metaphor, but what does it mean in reality?"

"Who knows what reality is."

"Good point."

A yellow-and-black rickshaw sputtered up and stopped in front of our house.

"Your coach has arrived," I said.

"Charming; then off to the ball."

We climbed in and drove toward Ahmedabad; children stood on the porches of their houses swirling sparklers in the air.

On the sand outside Goswami's house, a colorful tableau of Krishna and the milk maidens was painted on the ground in colored sand next to a large accumulation of sandals. The passionflower vines on either side of the door were in full bloom. Brendan slept in his carrier hung over Maggi's shoulder.

"Welcome, *namaste*," said Soma, embracing each of us as we passed into a room full of men in white shirts and dark slacks and women in colorful saris. Brendan woke up and peered out from his carrier with curiosity. Soma pinched his cheek. "Hello, cute fellow," he said.

Goswami emerged from the crowd and greeted us. "Happy Diwali, festival of lights; first we eat, then we play, then we talk." The racing crescendo of Ali Akhbar Khan's sarod, the Indian version of a guitar, poured out from a tape recorder on a table. Against one wall, a long table draped with garlands of marigolds was crowded with plates piled with crisp fried puff bread and mounds of purees, a deep dish of curried okra, a platter of cauliflower in cilantro sauce, and bottles of chutney.

"These are not very hot; we made them mild in your honor, very mild," said Goswami with a wink.

I dipped my finger in the curry sauce and lifted it to my mouth; the first taste howled on my lips. I reached for a glass of water and drank frantically.

"No, not hot at all," I said.

"All right, then, mix it with yogurt. Here." Goswami handed me a bowl of yogurt. "We shall never make you happy. Remember pain is in the brain."

"Yes, but hot is not," I replied.

We relaxed after dinner, listening to music. The guests sat on pillows scattered against the walls of the living room.

"Did you know every raga has its time of day?" asked Goswami.

"No," I replied.

"Yes, certain notes are in tune with certain times of the day. It is the goal of Indian music to lift the audience to a transcendent level, where the audience surpasses itself."

"What do you mean?" I asked.

"It's hard to explain, but it's as if the participants at a concert, musicians and audience alike, are lifted momentarily off the ground and float around in the air."

"I know the feeling." I thought of the rain dance at Woodstock, but thought better of bringing it up.

Later we climbed the hill behind Goswami's house. The silvery disk of the full moon in the east climbed with us as the slope of the hill slipped past directly in front of us.

"There's a interesting illusion." Goswami pointed to the moon.

"The moon's an illusion." I kept pace with him.

"The moon appears as a two-dimensional disk, yet we know it is a three-dimensional sphere." He stopped to catch his breath.

"I'm sure you are going to tell me why." I stopped beside him. Below us, a long line of men and women circled up the hill. Beyond, the city of Ahmedabad was lit up with rockets and fountains of light projected skyward.

"It's like this. The moon is covered with tiny mineral grains, dust really. The facets of the crystals act as mirrors that reflect the sun's light off the moon. Since the mirrors are arranged at random angles the same amount of light is reflected from the entire disk of the moon. Our brains are programmed such that they need shadows on the side to interpret an object as a sphere. At full moon, there are no shadows, so we see it as a disk even though we know better."

"Perhaps that is why it is so beautiful and mysterious."

"Perhaps."

When we reached the summit of the hill, the children ran and swarmed around, twirling sparklers in the air.

"Like little electrons moving so fast we can't catch them. By the time we look at them, they are already somewhere else. All we see is the trail of light," said Goswami.

We all sat in a circle on a flat meadow. A young girl came over and said, "Oh, baba, can we have him?" and she handed Brendan, sitting on my lap, a sparkler. Brendan kicked his legs, smiled, and looked up at me. He waved it in the air, and I watched that he didn't get it near his eyes.

The adults played drop the handkerchief, even women in saris and distinguished physicists. One person circled on the outside of the ring of participants and then dropped a handkerchief behind a person sitting on the ground. That person then jumped up and tried to beat the first person around the circle to the space. Whoever didn't make it became "it," and had to race against another person.

Goswami sat next to me; the children ran around the outside holding their sparklers aloft. "They are the electrons and we are the nucleus," said Goswami.

Soma's wife, wearing a blue sari, dropped the handkerchief behind Goswami. He was up and running in a flash, but she was already halfway around the circle. The men and woman howled with laughter, rocked on their seats, and clapped their hands.

"Is this seat taken?" She laughed and flopped on the ground next to me out of breath.

"Is this the way Indian physicists spend their spare time?" I asked.

"You mean playing? Yes, most of it," she answered.

We stayed in Ahmedabad another year, living in the same bungalow. After writing a second paper for the Deep Sea Drilling Project, I received notice of my appointment at Oregon State University.

I was packing up papers at my desk when Goswami entered the office. Brendan sat on the desk and played with a wooden puppet of Hanuman, the monkey god.

"Here, fresh off the press." I handed Goswami a copy of the Deep Sea Drilling Project paper.

"Oh, yes, thank you. Do you also have a reprint of your *Nature* paper?" he asked. "Autographed, of course."

"Absolutely, but this stuff is boring compared to particle physics."

"I suppose we all feel that fields other than our own are more interesting," said Goswami.

"What's this new idea about mini–black holes?" I asked. "That seems as interesting as anything could be."

"It's pretty much a theoretical idea but based on straightforward logic. Since the gravitation force increases as the distance between objects decreases, the very smallest units of matter might well be black holes so small that there could be up to a billion of them inside an electron. There could be more mini–black holes in a liter of water than there are stars in the universe."

"Why doesn't the intense gravitation force of the black holes cause the liter of water to collapse?" I asked. Brendan crawled from my desk to Goswami's. I reached for him.

"According to the idea of Princeton physicist John Wheeler, they are balanced by white holes that act in the opposite sense. Matter is destroyed in the black holes, but created in the white holes. Here, let him play." Goswami put on his surgeon's mask that he used in the fume hoods and made faces at Brendan, who giggled and reached for the mask.

"Is it the same matter recycled? I mean does matter enter a black hole and exit from an adjacent white hole?" I asked.

"Perhaps; the black holes and white holes are connected by wormholes, creating a seething foam at the smallest level of reality."

"And time does not exist in black holes?" I asked.

"Yes, at the center of black holes are singularities outside of space-time. The big bang expanded from a singularity. In fact, the quantum foam may be left over from the big bang. We are engulfed in the churning matrix of singularities outside of space-time." Goswami pulled a clean mask out of a drawer and fitted it on Brendan, who turned and made faces at me. He turned to Goswami and made faces back. Brendan rocked, kicked his legs, and they both laughed.

"So in a sense we are surrounded by a network of singularities outside space-time but connected by wormholes," I said. Here was a scientific explanation for a possible plane of existence outside of time. Perhaps Guruji was right; there is no absolute time.

"Do you think it is possible that thoughts interact with the quantum foam?" I speculated that the quantum foam is responsible for the connections in the Void that Guruji and Chombi spoke of.

"Perhaps. There is a school that suggests that thoughts are quantum waves that instantaneously travel through the universe," replied Goswami.

"Since the quantum foam is real, there is a deep reality. The world is not built on probability. Right?" I said.

"Right. According to David Bohm, the neorealist, reality is created by our thoughts projected onto the system of quantum waves arranged in a precise field of information, known as the quantum potential."

"Yes, the holographic universe. I know, but the problem with Bohm's theory is that in order to create reality, information must travel faster than the speed of light," I said.

"It's not so much that it travels faster than light, but that all points in the universe are connected through the system of wormholes and phaselocked quantum waves."

"That would seem to suggest that it is possible to see the future." I picked up Brendan and carried him to the window still wearing the surgeon's mask.

"Yes, if there is no time, then the future is now," replied Goswami.

"If we can see the future, doesn't that imply that the future is set, that there is no free will?"

"There is a new idea in quantum physics," replied Goswami. "Past, present, and future are set—but in different universes."

"You are talking about the many-worlds interpretation, right?" I asked.

"Right. In the many-worlds interpretation, the wave function does not collapse but continues to expand into its many aspects. The observer selects one aspect of the wave function to detect. All other forms of the wave function continue to exist but in parallel universes, where we are not aware of them. The multiple universe theory was first proposed by Hugh Everett, a graduate student of John Wheeler's at Princeton, in 1957. According to him, there is a super-reality that is a branching tree of possibilities in which everything that can happen, does happen. Each individual's experience is a tiny portion of a single branch on a lush and perpetually flowering tree."

"How does this splitting of universes come about? The idea is outrageous." At the same time, I had to admit it did solve the problem of free will in precognition. Each of the branching trails in Guruji's analogy could be a separate universe.

"True, but not any more so than the collapse of the wave function. In fact, the multiverse has advantages over the collapse of the wave function."

"Such as?" I asked.

"For one thing, critics of the theory can be shipped off to their own universe." Goswami laughed.

"Even so. . . ."

"Think of it this way. The multiple universe idea best explains the perfection of our universe."

"What do you mean?" I asked.

"We all admit that life is miraculous. But, if you understand physics, it is even more miraculous that the universe exists."

"For example?"

"For example, the gravitational force that pulls the universe together is perfectly balanced by the force that causes it to expand. If the force of gravity or the mass of the universe had been even a fraction higher, the universe would have collapsed. Stars would have collapsed into neutron stars and black holes. As finely tuned as the gravitational forces are, they are not the most amazing thing."

"What's that?"

"The smoothness. If in the early stages of the big bang there had been larger irregularities in the distribution of space and mass, galaxies and stars would have clumped together and blinked out of existence a long time ago. On the other hand, if there had not been some irregularities, we never would have stars or galaxies form at all. The smoothness of the universe is equivalent to that of a billiard ball with a few scratches."

"So the physics supports the idea that the universe was created just for man by a benevolent deity," I suggested.

"Or the universe is a supreme accident, a coincidence," he replied, "which is the standard scientific explanation. Until now."

"How does the multiple universes interpretation explain the perfection of the universe?" I asked.

"If our universe is just one of an infinity," replied Goswami, "we happen to live in the only one with conditions right for our existence. All the other universes are too hot, or too cold, or too dense, or too nebulous, or too short-lived to support life. By natural selection, we live in the universe that has the right conditions for the evolution of life. Darwinism on a galactic scale!"

"Is it possible that these universes expand from the singularities in the quantum foam?" I asked.

"Yes, it is possible that the quantum foam is the hatchery from which a plethora of universes hatch, but they are still connected together. The quantum foam is the basis of the super-reality that connects all universes," replied Goswami.

"If the multiple universe theory is correct, then we must experience a thousand different universes every second, one for each twitch of a peacock's tail," I said.

"True. Each universe is only a tad different from the previous one, imperceptible differences, really. Universes could be exploding, collapsing, replacing each other, one after the other, and we'd never notice. Thank God. There are some things I'd rather not know about," replied Goswami.

"Nonetheless, our brain presents us with a smooth, continuous flow of events."

"Just like our brains tell us the full moon is a disk." Goswami laughed.

"Good point. But we were talking about motion."

"The brain interprets a movie as continuous motion even when we know it isn't. Hey, brains are overrated. According to them, the earth is flat and motionless in space."

"So each universe could be like an image on the film."

"Precisely."

A peacock walked by on the lawn outside the window. "Look Brendan, a peacock. Can you say *peacock*?"

Brendan looked at me baffled and made a face under his surgeon's mask.

"Let's go meet him," I said.

"He'll just fly away," said Goswami.

"Shrödinger's peacock?" I laughed and carried Brendan outside.

I put Brendan on the grass, and he stumbled toward the peacock, grimacing, and dragging the Hanuman puppet by the strings. The peacock defiantly turned to face Brendan, rustled his feathers, and when he fanned his tail in grand array, Brendan stopped dead in his tracks. The peacock croaked softly, strutted forward, grabbed Hanuman by the strings and tugged on him.

"He wants Hanuman. It's okay; let him have him. I'll get you another," I said.

Brendan dropped the puppet. The peacock shook his beak, got a good grip, collapsed his tail, and, with a great commotion of flapping wings, flew off to the nearest mango tree.

"Everybody loves Hanuman. He's alive and well—in another universe," said Goswami, laughing.

Santabehn and her four children stood on the porch of our bungalow as the white Tata sedan pulled up. Soma jumped out and opened the trunk. Maggi came out of the house wearing a red sari and holding on to Brendan, who was now walking. Arjun and Mohandas helped me with our luggage. Soma slammed the trunk shut, and we gathered around the car.

"You have less going than you had coming," said Soma.

"Yes, no vitamins going," said Maggi.

"Don't neglect nutrition," I said.

"*Acha,* plenty in chapatis. I'm thinking of shipping several hundred pounds stateside. Chapati power, nothing like it, but you'll be back in Wonder-bread world soon enough. Then you'll be needing your vitamins," said Soma.

"Yes, and the way the world is going, everyone in India will soon be eating Wonder bread; then you'll be buying vitamins from us," I said.

"Makes you wonder, doesn't it?" joked Soma.

"Acha!" I said.

Veenu and Seeta piled garlands of marigolds over Brendan's head and pinched his cheeks. He smiled, laughed, and ran away from them on chubby legs. They chased him down, tackled him, carried him back to the car, and covered him with kisses while he giggled.

Santabehn placed a necklace of marigolds over my head, stood back, eyes glistening with tears, pressed her palms to her forehead, and said, "Good-bye, Sahib Roger."

"Good-bye, Santabehn; thank you; I will miss you." I pressed my palms together and bowed to her.

She placed a garland over Maggi's head, tears flowing freely now. Maggi burst into tears and pulled her into an embrace.

"We must be going," said Soma.

They waved to us as we pulled away from what had been our home for three years. We raced for the last time through the streets of Ahmedabad, past roadside stalls of okra and squash, and fried chickpea delicacies.

Guruji greeted us at the doorway of the ashram and placed garlands of marigolds over our heads. "Narayan, God bless you."

Several yoga students also placed marigold wreaths over our heads. Brendan's head barely peeked above the pile of marigolds on his shoulders.

Guruji took his place on the pillows below his portrait. We sat facing him.

"Guruji, I'm trying to understand these experiences we've been discussing for the past several years." Guruji's prediction of Brendan's birth had convinced me that at times in certain states of consciousness it is possible to tell the future. The only possible explanation was that somewhere in the universe time does not exist. He had also convinced me that some parts of our minds are connected to the universal quantum potential through thought. The best explanation is that thoughts themselves are events in the quantum potential. But I still had a few unanswered questions for him.

"Can experiences similar to my near-death experiences be had through religious practices?" I asked.

"True spirituality comes not from any organization or priesthood, but from the spiritual universe within," replied Guruji. "We live in a universe that is full of unconditional love and acceptance. Most important is knowledge and the ability to love. Learning is continuous and total knowledge is available in the afterlife; think of a question and you immediately know the answer. Knowledge is nourishment."

"Why is the quest for knowledge important during life if we have access to all knowledge after death?" I asked.

"It has to do with the ability of each individual to reach out and help others. In the afterlife, self-image is seen for what it is, a flimsy projection made real by the ego's desperate need to place importance on the time-bound mind and body. Spirit is a direct experience that transcends this world. It is pure silence teeming with infinite potential. All questions cease. Things are accepted, not judged; just being here in your body is the highest spiritual goal you can obtain."

"Our consciousness extends throughout the universe. Then it collapses to the material plane and we are cut off from the universe?" I asked.

"Correct, and in the expanded state, it is possible to alter the universe with thought," replied Guruji

"Only in the expanded state?" I asked.

"That is right. For example, telling the future is performed in a delicate state of mind and can't be done in the collapsed mental states full of anxiety, greed, hate, and lust. You must dissolve the ego or at least ignore it, and that is easiest done by giving up all previous perception of your personal reality. When you focus the mind on the higher levels, it merges with the energy of the Void. When you become self-conscious, or focus on the ego, you are cut off from the Void; that is why we can never measure its effects directly. We have to sneak up on the Void and can only detect it by not trying to detect it. As soon as you try, you are back in the illusion of materialism. This moment now is the heavenly moment. However, do not deceive yourself; do not confuse some fulfillment of greed or emotional state with the true bliss of enlightenment."

"Definitely. My near-death experiences taught me the value of losing the ego."

"And your near-death experiences taught you that you—all of us—are connected to the Void."

"That's true," I replied.

"In fact, your near-death experiences have taught you about life. What you do this moment is all that really matters, and you have a beautiful new son."

Guruji placed his hand on Maggi's forehead, then Brendan's, then mine.

"Do not despair," he continued. "Enlightenment will find you when you least expect it." He smiled with a great light in his eyes that filled our hearts with love. "I will remember you always. Remember, by wanting something, you change it; if you love something sincerely, it will erupt spontaneously from the Void. You have more power than you believe. You are not a victim of the world, forced to accept conflicts, but are free to choose from an infinity of wonderful possibilities."

"We, too, shall remember you always," I said.

We bowed before him and left the ashram with Soma.

Once in the car I said, "Let's swing by Gandhi's ashram on the way to the train station."

"You haven't been there yet?" asked Soma.

"No. Is there time?"

"Let's go."

Crouched together directly on the high bank of the Sabamarti River, a series of simple outbuildings with red-tile roofs and broad porches had been turned into a museum with paintings and relics from the life of the father of India.

"This painting depicts the beginning of Gandhiji's famous salt march. He broke English law by making salt on the banks of the Arabian Sea. He spent half his life in jail and survived two terminal fasts to prevent widespread rioting and bloodshed," said Soma.

"Why so many spinning wheels?" I was amazed by the collection of wheels for making homespun cloth.

"It was a political statement. Gandhi believed foreign industry caused Indian poverty. He advocated decentralized, low-capital, cottage industries to break the syndrome of poverty. He wanted everybody to produce their own food, clothes, and shelter."

"Did it work out?'

"The idea was never actually tested. After his assassination, foreign capital investment in industry became more prevalent than ever. Cottage industry was stamped out, and poor Indians were forced to leave their villages and live in poverty in slums surrounding industrial centers where they could sometimes find work."

We strolled to the bank overlooking the Sabarmati River. The dry season had begun again and the riverbed was mainly bare sand. Women beat clothes in the shallow water with sticks. Boys sprayed water on their buffaloes. Miles and miles of freshly dyed red cloth draped over drying sticks hung along the bank. Two sadhus with a white cow crossed on the opposite side.

"Look!" Maggi said. "It's the two sadhus."

"Oh," said Soma, "they're charlatans; ignore them."

We drove slowly through crowded streets toward the train station. Crowds of people milled every which way, oblivious of the blare of our driver's horn. Shopkeepers sat cross-legged in front of stalls selling dry goods, textiles, and silver jewelry. The station itself was in the shadow of textile factories belching clouds of steam and vapors into the air.

The locomotive engine let out spurts of steam from the driving wheel cylinders, and a steady plume of coal smoke curled skyward from the smokestack. Three porters carried our luggage back to the first-class car using trump-lines over their foreheads. A small fan circulated the air above the cushioned seats with linen-covered headrests.

"This is it. At last, you are traveling first class," said Soma.

"We can't really travel third class with Brendan," replied Maggi. She set Brendan on the seat with a new wooden string puppet of Hanuman to play with.

The train gave a jerk as the engineer released the brake locks.

"It looks like this is it," said Soma.

"Thank you so much, Soma; we are very grateful," said Maggi.

"I will come and visit you in Oregon next time I'm at Scripps." He kissed Maggi on the cheek.

"Well, I'm not going to kiss you, but safe journey." He clasped his hands together and bowed slightly.

"Thanks, Soma. We'll be seeing you stateside," I said.

He bounded down the steps to the platform as the train started to move. We watched him pass from view as the platform sped past us. The engine got up to speed in jerks and then settled into a steady rhythm as the streets of the city swept from view. Open fields and cactus hedgerows moved past as the train gained speed. Herds of camels and water buffaloes floated by, and after switching to the tracks bound southward to Bombay, the train crossed a bridge over the Sabarmati River and the Gujarat Desert spun past. The vegetation took on a tropical cast, and hills rose up to the Deccan Plateau. Women labored in ankle-deep water between rows of rice that flashed by with the strobelike precision of spinning bicycle spokes. The sun set over the Arabian Sea; the lights came on in our cabin. We flashed through a tunnel as though it were a wormhole to another universe, and came out the other side with the cabin's lights flickering. They went out for an instant, and we saw the sky cast with red. The lights came on again, and the windows went black, reflecting the contents of our cabin.

CHAPTER ELEVEN

INFINITE FUTURES

Shortly after we moved to Oregon, in 1977, I experienced another *eureka moment* in my professional career. The moment came one night while I was walking through the campus of Oregon State University. For an instant my mind was tuned to a deep level of understanding. I had been studying the chemistry of volcanic rocks dredged from the bottom of the ocean. When volcanic lava cools rapidly, it forms a glassy rind, *obsidian,* on the outside that traps the gas components carried by the erupting magma. Certain compositions and isotope ratios in these glassy rinds are identical to those of gases in the earth's atmosphere.

At first I thought the atmosphere had contaminated the rocks. But for a number of reasons this was not possible. The eureka moment came when I realized that the earth's atmosphere had formed by release of magma from the earth's interior, leaving pockets of gas behind! Naturally, this residual gas would bear a resemblance to atmosphere gas, because, in essence, they were the same. I published my conclusions in *Nature,* in 1979, followed by another article in the same scientific journal in 1983.

Science is like climbing a mountain: the higher we go, the farther we see. Each new view yields a greater understanding of the interconnectedness of things. On the summit we attain a new moment of understanding as new observations are integrated into the landscape of the prevalent worldview.

I am addicted to the moment of understanding, to the satisfaction of scientific discovery. When a new idea, a new framework of how things really are, is born, the rush lasts a lifetime.

What is the moment of understanding, and how does it happen? For me, it begins with a slow gestation of observations in the womb of the unconscious. I become totally familiar with the data of observations, their correlations and configurations. Then I forget them. Usually when I least expect it, a flash of intuitive insight emerges into consciousness. It is that moment of realization that exhilarates me. These eureka moments had a lot in common with the near-death experiences; normal thought processes had ceased. I speculated that the near-death experiences facilitated eureka moments by opening channels to the unconscious mind.

Because of their unpredictability, I was unable to design research programs that would specifically lead to eureka moments. Yet I was unsatisfied with anything less. Writing research grants, outlining scientific proposals, and dealing with administrative bureaucracy stifled the mental activity that lead to the eureka moments.

I waited for my chance to analyze my May 29 experiences, in the hope that they would lead to new eureka moments. What new moment of understanding awaited me? The questions consumed me: Does consciousness survive death? Is there a reality independent of the observer? Does time stand still in the heart of the universe?

The May 29 experiences shattered the reality I had inherited growing up in middle-class America; the linear suburbs were built on laws of geometry and trigonometry over a millennium old and fed by the shallow promise of predictability in social and technological organization. The abstract map of the terrain had replaced the terrain itself. But I did not realize this until I visited places where there are no maps—winding cobble streets of Andean villages, labyrinths of Moroccan medinas, and the complex architecture of high Himalayan passes. Although frightening and challenging at first, the places without maps stimulated, nourished, and awakened levels of unconsciousness that had been veiled by a conscious mind that fought to preserve its version of reality at any cost.

Adventures of physical hardship in remote lands evolved into experiments in the way the human mind interprets the world. Reality changed with locality. The worldviews I encountered were only partial views. The question arose: Is it possible to compile a complete worldview from partial experiences?

I decided to analyze the May 29 experiences as if they were scientific experiments, to find out if, by deduction, I could infer a model of deep reality that was universal, internally consistent, and, perhaps, testable by further observation.

But first I needed to reconsider the scientific process, specifically the dictum of objectivity. Objectivity has been the cornerstone of science, that which gives it moral supremacy over religion, but science itself has shown that the object of study cannot be separated from the observer. At the grand scale of the universe, Einstein showed that at velocities approaching the speed of light, space shrinks and mass becomes infinite. At the smallest scale of the universe, a world of quantum waves rippling on an infinite sea of energy turns into matter when we touch it with our perceptions. The observer's choice of what to look for has an inescapable consequence for what is found.

No matter how objective we are, can we see things as they really are? Research has shown that our conscious view of reality is a simulation built on less than one part in a million of the total information that enters our minds. We do not experience the world as raw data. We become aware of an experience only after it has occurred.

Furthermore, the reductive, materialist worldview that dominates Western science cannot solve the complex problems of patterns that emerge from nature at all levels of reality. We cannot describe the world without including in our description the fact that we are describing it. Since I could not isolate my mental state, feelings, physical surroundings, the influence of my genetic code, or the manner in which my brain processes data, I included my prejudices, conditionings, and worldview, precisely and accurately, as part of the experimental setup and inferred those conclusions I could honestly make.

I recorded my observations and drew my conclusions without subscribing to a preconceived view of reality. In fact, my model of reality continuously evolved during the process of making observations.

The May 29 experiences revealed that when normal brain functions are subdued, a heightened state of awareness emerges from unconsciousness. Cessation of normal thought is accompanied by time dilation and euphoria. The experiences broke down into three results: (1) mind is separate from body, (2) reality is interconnected, and (3) time is observer-created.

I experienced the first result of the May 29 experiences, the independence of mind from body, during my fall on Mount Everest; my awareness separated from my body as it careened down the ice cliffs. I concluded that the human mind or a part of it precedes and survives the body. The existence of an immortal mind explains a whole range of observations, but cannot be measured or proved directly. On the other hand, I could find no proof that mind, in the sense of pure awareness, does not survive independent of the body.

The concept of mind-body dualism was first proposed by René Descartes, who argued that he could imagine his mind existing separately from his body, so his mind could not be identical to his body. On the other hand, many scientists assert that mind is the result of a combination of physical processes in the brain. After all, different parts of the mind are affected when parts of the brain are damaged. Brain activity can be mapped out by modern imaging equipment and correlated to specific mental states and thought processes. Changing the neurotransmitter ecology of the brain can alter thought patterns. These observations show that the brain influences thought processes and mental states; however, just as TV programs do not originate in the television set, the mind does not originate in the brain. The idea that mind exists apart from the brain is corroborated by simple considerations of brain physiology and neurochemistry that show physical attributes cannot account for all aspects of consciousness.

First of all, there is not enough room in the brain, or in any other place in the body, to store the wealth of information contained in the average mind. John von Neumann placed the storage capacity of the mind at almost a trillion gigabits, over a million times the storage capacity of all the computers on earth.

The brain contains about a trillion cells, each of which may have up to a hundred thousand connections. But these numbers pale in comparison to the storage capacity of the average mind. In order for the mind to be located exclusively in the brain, each connection must be involved, on the average, in the storage of ten thousand bits of information. Therefore, the storage capacity of the human mind, if it is physical, must involve circuits a million times smaller than the fastest computer microchip, at the scale of the atom, at the quantum level.

To make biological functions possible our consciousness filters out, discards, most of the information in the mind, so that it won't overwhelm or confuse us. For every bit of information brought into consciousness by the brain, a million bits are filtered out. The brain blocks out vast quantities of information that normal consciousness cannot handle.

Furthermore, the unconscious mind responds to stimulus before the conscious mind does. The brain makes us aware of our perceptions *after* they occur. Nerve cells work electrically. Measurements on the electric fields of the brain, through EEG (electroencephalograph) machines, show that an electric current precedes physical acts by about a second before the brain processes them at the conscious level. Our unconscious mind prepares for any action about a second before we are conscious of the act! We are the last to know our thoughts.

Even the simple act of watching a movie requires a million neurons to fire simultaneously. How is all this activity coordinated? Visual information entering the eye is scrambled so that the right side of the field of vision ends up in the left half of the brain and vice versa. A precise matchup must be maintained to ensure that both eyes see the same thing at the same time. The sound track is registered through tiny hairs in the auditory canal, each one vibrating at a different frequency and stimulating an electric nerve impulse. Science has yet to discover differences in the nerve impulses produced. Yet the sound track is matched perfectly with the deconstructed visual signals and projected simultaneously onto the screen of consciousness.

The sound and visual tracks of a movie are precisely mixed by computer editing programs. An as-of-yet-undetected quantum timing mechanism must match up the auditory and visual signals of the brain.

There are many components to consciousness, and some of them—such as awakeness, introspection, reportability, self-consciousness, voluntary control, and attention—can possibly emerge from material sources in the brain. However, there are other aspects of consciousness that cannot have roots on the material level. Sir Isaac Newton referred to them as phantasms of the brain. Recently they have become known as qualia phenomenon, or the experience of realizing we are experiencing something. The common feature of all these experiences is that they are not stimulated by a specific physical object. There are no nerves whose function is to let us know we are experiencing something.

The experiences of moral judgment, of pain, of hot or cold, of beauty, of any emotion, of negative facts, are all possible phantasms of the brain. This list of possibilities is by no means exhaustive.

Let's consider the experience of color as an example. On first thought, it seems that color does have a physical origin in a specific wavelength of light. However, after light impacts the retina at the back of the eye, the frequency information is lost in the conversion to electrical nerve signals. The nerve signals generated by red light are the same as those generated by blue light. Some colors, such as purple, do not correspond to a particular frequency in the first place.

The impression of color is very much a subjective experience, as illustrated in the following thought experiment. In a dimly lit basement Mary washes her white sheets with a blue sock. The sock runs and tints her sheets a faint hue of blue. When Mary hangs up her sheets, her brain, used to perceiving sheets as white, makes an automatic adjustment so that she sees them as white. Tom brings his white sheets to the basement, and when he hangs them up next to Mary's, they appear pink to her because her brain has compensated by applying a pink filter to her perception. Tom sees Mary's sheets as blue because his brain has made no adjustments.

The philosopher Thomas Nagel illustrated the ephemeral nature of the experience of color with the following thought experiment: Imagine we are living in an age when we know everything that there is to know about how the brain generates human behavior. Mary is the world's leading neurophysiologist, and she can make a robot with the same exact optical-perception apparatus as a human.

Imagine Mary has been color-blind since birth. She has never seen color, only black, white, and shades of gray. She cannot physically hardwire her robot to have the experience of color because the experience is not inherent in physical systems. The robot cannot start seeing color unless Mary points to colored objects and names the color: That's blue. That's red. She cannot teach the robot because she has no experience of color herself.

The same thought experiment applies to other abstract qualia that are not inherent in physiological systems. A nonmaterial mind must be postulated to account for our perception of qualia.

Separation of the mind from the body is universal in accounts of near-death experiences. Even blind people, clinically dead, with flat-lined EEGs, have accurately described the clothing and jewelry that surgeons and nurses wore in the operating room. Over ten million Americans have had near-death experiences; it is no longer possible to dismiss them as aberrations of individual experience. According to Guruji, it is the death of the ego, of self-consciousness, that leads to the experience of a mind separate from the body, and this may be the basis of all transcendental experiences.

There is also the problem of the screen of the mind. Science has mapped out parts of the brain that receive specific stimuli or generate specific responses. These perceptions are presented to us as if they are projected onto a computer screen. The screen of the mind must receive coordinated perceptions from all parts of the brain. So far, the location of the screen of the mind has eluded all attempts to find it within the physiology of the brain. The screen of the conscious mind is refreshed every half second or so. This means that we really experience the world as a series of still images, like individual frames of a movie. Yet the single frames produce the illusion of a continuous experience. How does the mind register disparate signals to present us with a continuous experience? Given the limitations of brain physiology, it seems likely that the screen of the mind is nonmaterial.

The hypothesis of separate mind and body follows a simple analogy from quantum physics. Matter exists in two forms, the wave and the particle. The mind corresponds to the wave function of the quantum world, the body to the particle phase. Most of the time we focus on a simulation of reality in the collapsed particle phase. During the near-death experience we abandon normal processes of consciousness and participate directly in a deep reality through the quantum states that make up our nonmaterial minds.

If John von Neumann and Eugene Wigner were right, consciousness collapses the wave function. Yet the consciousness that collapses the wave function cannot be produced by a material brain that is already collapsed. What collapsed the matter of the brain? The argument continues until a nonmaterial, uncollapsed, state of mind is encountered. A nonmaterial mind must collapse the wave function before the brain experiences it.

So, how does the nonmaterial mind interact with the brain? The best explanation is that the brain acts as an antenna, a receiving station for quantum waves of the nonmaterial mind. Several possible links have been suggested. Perhaps, the nonmaterial mind, at the quantum level, stimulates release of neurotransmitters into the gaps between nerve cells. It is possible that the tiny hollow tubes, *microtubules,* that make up the skeleton of all cells resonate with quantum waves of the nonmaterial mind. The quantum forms of DNA, RNA, and other molecules may be connected to quantum aspects of nonmaterial mind through a common substrate.

My reading and thinking about the mind-body problem convinced me that mind-body dualism is a reality and that my experience of a nonmaterial mind was not a fluke. Once I was able to fully accept that my mind did really separate from my body on Mount Everest, it became possible to analyze other aspects of the May 29 experiences and postulate reasonable explanations for them.

The second result of my May 29 experiences, one not described in most near-death experiences, is the realization that thought influences reality. The ability of thought to change reality emerged in three extraordinary events: (1) the blizzard in Tierra del Fuego stopped just after we made the visualization, (2) the sun came out during the rain dance at Woodstock, and (3) the car crash in Morocco occurred when Maggi and I pleaded with Hussein to let us out of his car. By themselves, I would have dismissed any one of these events as coincidence, but three events were beyond the probability of chance.

If any one of these events had occurred to me alone I would have dismissed it as a temporary disorder of the senses. But Peter, Paul, and Jack agreed something extraordinary took place during the blizzard in Tierra del Fuego. Several thousand participants can be seen rejoicing at the appearance of the sun in the movie *Woodstock*; and Maggi confirmed my recollection of the accident in Morocco.

All three events happened when normal thought processes had ceased. In Tierra del Fuego and Morocco, I had been convinced I was about to die. At Woodstock the dancing and drumming induced a trancelike state.

My conclusion is that mind and matter are independently connected to a common substrate, we are blocked from this connection by the normal processes of consciousness. In the near-death state, the block is removed and the connection is experienced. If the mind exists independent of the body, it could exist in the quantum world where it might interact with quantum waves.

I had, and still have, difficulty accepting this result. I was taught at a young age that only God has sanction to influence reality, that I do not possess special power, and that to think so is sinful and egomaniacal. Even if it is possible to change reality with thought, perhaps we should not do so because it is confusing.

Nonetheless, I have experiential knowledge that it is possible to change reality with thought. One confirmed event of changing reality is cause enough to constrain our ideas on the nature of deep reality and human thought.

The interconnected (nonlocal) nature of the universe is confirmed by the common occurrence of those unusual and meaningful coincidences that Carl Jung named synchronicities. Because they can hardly be attributed to chance, he proposed that synchronicities are evidence of an acausal connecting principle unknown to science. What are the chances that the most meaningful experiences of my life, those marked with euphoria, time dilation, and separation of the mind from the body, would occur on the same day of the year, May 29? Like me, most people have experienced extraordinary synchronicities at one time or another, but have dismissed them (as I first did) as aberrations. Nonetheless, synchronicities are evidence of a hidden order shared by mind and matter. Extraordinary coincidences and synchronicities are momentary fissures in the fabric of space-time that allow glimpses of immense underlying order.

Quantum mechanics offers an explanation for the interconnectedness of the universe. It is caused by the phaselock of quantum waves in the information field of the quantum potential. Two photons with correlated attributes traveling in opposite directions instantly respond to changes in one or the other. This is because quantum waves have attributes that resonate with each other, like fireflies flashing in time with each other, or an unplucked guitar string that sings when another is plucked in harmony, or the pendulums of wall clocks that swing in unison. The feedback information that locks the lighting of fireflies is provided by visual stimulation, while that of wall clocks and guitar strings is provided by molecules of air. There must be a source of undetected feedback information that brings quantum waves into phaselock, a common field of information in the quantum vacuum. Perhaps the mind links to quantum waves because *it* consists of quantum waves. Through feedback loops, the mind might come into resonance with the vibration patterns underlying the material world.

The universe must be interconnected (nonlocal) and linked to the human mind through a field of information. How does that specifically enable us to change reality? I decided to employ the method of multiple working hypotheses, wherein all possible explanations, even if they are outrageous, are considered and a series of tests and arguments are applied to decide between them.

The first hypothesis is that the ability to change reality invoked the existence of a special mental "force." According to the "force" model, some energy emanated from us in Tierra del Fuego and physically parted the clouds. The dancers at Woodstock conjured up a group "force" that caused the sun to come out. A "force" field must have passed from me to Hussein causing him to swerve at the wrong time, thus causing the wreck in Morocco.

But science has never detected such a force field.

There is a less dramatic but more outrageous hypothesis that may account for changes of reality by thought. What is reality anyway? The reality of a frog is different from that of a dolphin because their brains process information (electrical signals from their sense organs) differently. A Pygmy raised in the rain forest experiences a different reality from the one of an Inuit raised on the Arctic tundra because their brains interpret perceptions differently. Even within individuals of the same race brought up in the same environment, reality changes with brain chemistry and traumatic experience.

Our brains simulate reality by processing and discarding information, just as a statue emerges from stone when the sculptor chisels away fragments. Changing reality may be like changing computer programs of virtual reality on the holodeck of the Starship *Enterprise*. Our brains are the computer capable of interpreting information according to various programs.

According to this hypothesis, our brains' simulation of reality from the quantum field of information changed during the blizzard in Tierra del Fuego. The computer program that created a holodeck representation of a blizzard was replaced by a program that included a brief cessation of the storm. The rain stopped at Woodstock because the brains of the dancers were reprogrammed to include the sun. The program in my brain changed twice during the car crash in Morocco: first, when the crash occurred; second, when we all survived. The difficulty with this hypothesis, or any other that attempts to explain a change of reality by reprogramming of the brain, is this: How can disparate individuals experience the same change of reality? That question perplexed me for some time.

The third result I derived from the May 29 experiences is that precognition, the ability to know the future, is possible. I knew I was going to run into Maggi in the movie theater, and Guruji's prediction of Brendan's birth suggested it was possible to know the future. Even one reliable instance of precognition requires a new concept of reality. If the future is knowable in the present, then the apparent separation of the future from the present by time must be an illusion. At some level of the universe, time must not exist. When Guruji predicted Brendan's birth, I was forced to accept the possibility of precognition. Nonetheless, I would have trouble believing precognition even if it was proved true beyond a shadow of a doubt unless there was a satisfactory scientific explanation.

Philosophers have elucidated the paradox of time since the beginning of time. Time is the measurement of the progression of events. Yet what preceded the first event? And what preceded that one? And so on. The only way out of the paradox is to assume that there is a prime mover that is eternal, outside of time. Hence, there is someplace (or someone) in the universe where time does not exist.

Science is just beginning to consider the problem of time, and it, too, like philosophy, is beginning to reason that there must be an eternal framework beyond time. It is generally assumed now that time does not exist in singularities in the core of black holes. Since our universe evolved from a singularity, time (as well as space) began with the big bang.

The scientific reasoning follows from the uncertainty principle of quantum physics. Certain complementary attributes of nature always add up to the same quantity. If one value is high, the other is low and vice versa.

Time and energy are complementary. Enormous energies are possible for very short times, on the order of a billionth of a second or less. Conversely, within enormous concentrations of energy, and that is, mass, such as a black hole—time is extinguished like a blown-out flame. Similarly, as objects increase in mass and therefore energy as they approach the speed of light, accompanying clocks slow down, and time stops.

The two greatest theories of the twentieth century, quantum theory and general relativity, postulate that time is a secondary concept. It has meaning only on a scale at least a billion times larger than the quantum world and well away from black holes and singularities. Science no longer postulates that time is a river that rolls inexorably forward or is an absolute frame of reference. If time is not absolute, then it is possible that the future and the present exist side by side and that knowledge of the future is accessible in the present.

If time does not exist, why are we late for appointments? Why do we grow old? Our consciousness must construct everyday, time-bound reality from an order of existence where time does not exist. During near-death experiences, the processes that join futures together in a sequence cease, and time stops. Once we become aware that we have chosen a future, time starts up again, and we lose contact with that part of us that creates time. It happens so fast we don't realize we are doing it. Although this hypothesis is compelling, it is at odds with our normal worldview.

There is the additional problem that precognition appears to violate the concept of free will. In order for us to see the future, it must be fixed. Does the ability to see the future imply that we have no free will—that our futures are predetermined? That question perplexed me ever since the meeting with Maggi in the Eighth Street Cinema.

I was pondering this question when the first eureka moment of my analysis of the May 29 events came. I was driving to my office at the Hatfield Marine Science Center, along highway 101 near Newport, Oregon. Even though the day was overcast, the sun reflected off the shimmering water with a silver aura. Six pelicans rode the air wedge in front of large wave before it broke. Some of my best thinking has taken place while driving a car. The task frees my mind because it doesn't have much to do, other than follow the rules of the road. The eureka moment was the realization that we have free will and can see the future because we ourselves choose one future from an infinity of possibilities. We have the free will to experience one future without altering the others. Seeing the future is really the same as choosing the future.

We see other people's futures by choosing to experience the future in which they match our precognition. They continue to exist in an infinity of other futures, an infinity of other states not experienced by us.

Just as the resolution of our optical perceptions is inadequate to directly perceive the profound emptiness of a brick, our brains fail to resolve more than one future at a time. Time is created by experiencing a sequence of static futures, one at a time, like experiencing the static frames of a movie as continuous motion.

There was an added element of realization in this *eureka moment* that changed my view of life. My scientific training had taught me that for every bit of energy used up there is a percentage lost as entropy that can never be recovered. This pessimistic philosophy stemmed from the second law of thermodynamics, which was developed to explain the energy balance of steam engines in the nineteenth century. Applied to the cosmos as a whole, the second law of thermodynamics predicts that the universe will eventually dissipate all its energy in a kind of heat death. But if we are indeed creating time by selecting futures from an infinity of possibilities, then we can enjoy life with optimism. The outrageous hypothesis of infinite futures means anything is possible.

How can an infinity of futures exist concurrently? How is it that we choose futures, and why are we unaware that we are doing so? These questions remained unanswered by the outrageous hypothesis. Yet it exhilarated me because it solved the one outstanding problem posed by precognition, that of free will.

The sequence of understanding derived from my May 29 experiences followed in reasonable steps. The human mind exists independent of the body; it can change reality with thought and creates time by selecting futures from infinite possibilities.

John Wheeler's quantum foam idea offered a reasonable model to support my conclusions. At the core of matter and space all around and within us, so small that we cannot detect them, the world is made of mini–black holes and mini–white holes bubbling out of the quantum potential energy, forming and dissolving by the trillions within every elementary particle. The black holes are balanced out by the white holes, creating an equalized pressure that we cannot detect. Matter and time collapse into the black holes and emerge out of the white holes. At the center of every mini–black hole and every mini–white hole is a singularity, where time stops. We are surrounded by and permeated by a fabric of reality that is beyond space and time.

Interconnectedness is maintained by a network of wormholes that instantaneously transmit quantum waves throughout the universe. Deep reality itself is nonlocal.

The human mind, or at least part of it, is directly in contact with the quantum world because it is part of the quantum world. Perhaps some types of thoughts are quantum waves. All possible futures exist concurrently in the timeless matrix of singularities; mind chooses the futures that manifest as personal experience. Since the consciousness created by the human brain cannot experience everything at once, it experiences a sequence of futures creating the illusion of time. At some level we have the free will to choose between possible futures. Seeing the future is really choosing the future; other futures still exist and continue on. The precognition that I would run into Maggi in the Eighth Street Cinema was really choosing to experience that one future among infinite possibilities.

The navigators in Frank Herbert's science fiction classic *Dune* travel to distant parts of the universe by consciously choosing a sequence of futures that lead to the presence of themselves and their passengers at the designated location. The navigators in *Dune* develop conscious control over which futures to experience by ingesting a magical spice. In 1957, Hugh Everett proposed that the wave function does not collapse but continues through time in its many aspects. The observer splits into many different observers, each one experiencing a different aspect of the wave function. According to this idea, each of us splits off millions of observers every day, but we are only aware of one, just as a person suffering from multiple personality disorder is aware of only one personality at a time. Under normal circumstances, we have no control over which observer to instill with consciousness.

Under extraordinary circumstances, like the navigators in *Dune,* we can, through the quantum aspects of our minds, choose which observer to experience. According to Guruji, love is the spice that allows the choice; love focuses the desires on a particular observer in a particular branch of reality. In the near-death state, we lose the focus of consciousness through the ego and become aware of ourselves as observers in other possible futures.

My analysis had answered some of my questions. The answers were outrageous, and I could hardly believe them, but new and more perplexing questions arose. If time is an illusion that we create, is there a deep reality independent of the observer? Why and how do we collapse the wave function, create time, and transform an interconnected universe into one broken up into local parts? If we each simulate our own realities, why do we seem to share common experiences, or each other? The mystery deepened.

Unless, of course, I was willing to discredit the May 29 experiences as figments of my imagination.

It was May 28, 1980. We were camped on Mary's Peak in Oregon. Across the Willamette Valley the full moon hovered above the snowcapped stratovolcanoes of the Cascade Range. The night was clear and still. The sparkling gems of streetlights traced out the avenues of Corvallis below us. I was expecting Paul Dix, my friend from the Tierra del Fuego blizzard. A burst of sparks jumped up from the campfire as Brendan poked it with a stick.

"Grilled to perfection!" Maggi joked as Kieran, age three, pulled his flaming marshmallow out of the fire. "Marshmallow flambeau, *blanco y negro*."

"Here, Kieran, let me extinguish that for you." I blew it out.

"I like it like that." Kieran stuffed the marshmallow in his mouth, already covered with enough white froth to frighten a rabid dog.

"Watch this. Guess what I'm drawing." Brendan, who would be six the following day, pulled his orange glowing stick out of the fire and scribbled in thin air with it.

"A marshmallow." Ian, age five, looked up from his roasting chores.

"Nope." Brendan repeated the gesture, this time in cooler cherry red.

"The moon." Maggi stuck another marshmallow on Kieran's stick.

"Nope."

"A cookie," guessed Kieran.

"Nope. It's a *B* for Brendan." The stick went black, and he poked it back into the fire.

"Oh, no wonder. You have to write it backwards for us to read it," I said.

"Time for bed." Maggi rolled out the bedding for our sons in the tent.

"They'll be sugar-hyped for hours," I said.

"They can watch the fire from the door of the tent." We hugged each of them before they crawled into the tent. The moon, now silvery, shone brilliantly through the dark prayerful silhouettes of Douglas fir boughs. Not a needle rustled. The smoke from the fire rose straight up.

"By the way, thanks again," I said, pointing to the children in the tent.

"For what?"

"For giving birth to our children."

"Sure, my pleasure." Maggi sat down on a cot next to the fire and rearranged the embers with Brendan's stick. "Almost May twenty-ninth; do you ever feel you should spend the day in bed?"

"I think the May twenty-ninth experiences are over." I sat down next to her and covered us with a blanket.

"Why's that?" she asked.

"I'm not sure, but I think it's because I think about them too much."

"Don't you believe thinking creates reality?" Maggi blew on the end of the stick and watched the sparks fall back into the fire.

"Not so much *thinking* as *choosing*." I reached down and threw another log on the fire. A burst of sparks swirled up, illuminated the canopy of fir branches overhead, and died out.

"I choose; therefore I am." She laughed.

"I think so." I laughed too and smiled at her.

"Seriously, do you think we choose the future?" she asked.

"I think some part of us chooses the future from an infinity of possibilities. That's my hypothesis, but I don't really believe it, not at the molecular level," I replied. "Now, Woody Sayre lived it. When he said anything was possible, it was from firsthand experience. He felt it in his bones."

"Why don't you believe it like that?"

"I believe it at the intellectual level," I replied, "but my consciousness doesn't believe it at the experiential level of hardwired neurons and dendrites." I pushed a log into the fire with my toe.

"Wouldn't it be great if we could bring up our sons able to see the world as it is, full of possibilities?" Maggi poked the tip of Brendan's stick back in the fire.

"It would be great, but probably impossible," I replied. "Who knows what will condition them? Accidents happen. At best, we can limit our own conditioning and set an example."

We fell silent and stared into the fire. I held her hand and caressed the wedding ring I had made for her. We had carved the wax casts ourselves. A dentist friend cast the rings in gold. She carved a yin-yang symbol on mine. I carved the full moon shining through Douglas fir branches on hers.

"Look, the ring matches the night." I held her hand up against the sky so she could see the moon in the fir branches. We had fallen in love with the temperate rain forest of the Pacific Northwest, the lush vegetation, thick ferns, lichens and mosses on every tree branch. We had decided to stay in Oregon, even though it was a backwater as far as establishing a career in science. We roamed the dense forest and surf-washed shores of the Pacific every weekend. Mary's Peak was one of our favorite hangouts—we camped there on Brendan's birthday and several other times every year. The headlights of a car bounced through the tops of the trees below us.

I stood up to watch the progress of the headlights up the mountain road. Paul and his friend Robin Jahr were driving from Montana to San Francisco and had promised to stop for a visit. Paul and I had spent several years traveling around South America doing photography after our adventure in Tierra del Fuego

"So your May twenty-ninth experiences are over. What do you make of them?" Maggi asked.

"Each adventure taught me something." I sat down next to her.

"The first, Mount Everest?" She asked.

"On Mount Everest I learned not to be afraid of death, that

awareness of awareness continues after death. Woody taught me anything is possible. Chombi taught that the everyday world is an illusion. But I learned that illusion is only a partial answer."

"And Morocco? I remember that you had visions of car crashes before the accident happened."

"I was going crazy, totally confused, shifting from reality to reality without hope of returning to the correct one, not even sure there is a correct one. Then I focused on the prayer from *The Tibetan Book of the Dead,* and everybody miraculously survived the accident."

"Yes, I remember you thought you had died and that I was your anima. I had to assure you that I was real."

"Then there was Guruji's prediction of Brendan's birth. I do trust Guruji. What if any of the millions of premonitions that happen every day to millions of people really happens?"

"Some of them must really happen." Maggi turned to me, the fire danced in her eyes.

"Then the future, or at least part of it, is present now." I said. "Time must be an illusion."

"But time does exist." She stared at me in quiet expectation.

"Sure, but we create it. We extract time and space from a timeless, infinite foundation of the universe."

"The word of God, as they say in the Bible."

"The word, the mind of God, whatever you want to call it."

"If it is possible to tell the future," she said, "our lives are predetermined; we have no free will."

"That is where the many-worlds interpretation enters. Billions of parallel universes exist simultaneously; every possible universe that can exist, does exist. And new ones pop up every time we make an observation—along with a new version of ourselves that we are not aware of."

"What does that have to do with knowing the future?"

"Knowing the future becomes the same thing as choosing a future."

The headlights of a VW microbus rounded the bend on the road below. Wind rustled the tops of the trees and wafted the column of smoke into our faces. Paul and Robin pulled up in their van and parked next to our car. Their headlights drowned out the light of the campfire and reflected on the column of smoke and the dusty tree trunks. I jumped up and ran to the van. The lights went off and the light from the fireplace flared up.

"Hola, Pablo." I gave him a hug as he emerged from the van.

"Roger, good to see you." Paul's blue eyes sparkled as bright as ever.

Maggi hugged Robin, then Paul.

Robin was born in San Francisco of Japanese parents. Long black hair flowed over her shoulders and amused eyes peaked out from under well-trimmed bangs.

"Good to see you guys." I pointed to the cot near the fire pit.

"Likewise," replied Robin.

"We expected you earlier," Maggi said.

"We were late getting on the road," said Paul. "We've been going twelve hours."

"We had hot dogs and marshmallows for supper, can I interest you?" asked Maggi. "Kids, you know."

"Sure, but no thanks, We stopped for couscous and black bean sauce on the Columbia River," said Paul. "Man, those waterfalls are great."

"How about some wine, then?" I motioned for them to sit. Maggi and I sat on a log beside them.

"Roger and I were just talking about his May twenty-ninth experiences," said Maggi.

"The May twenty-ninth experiences?" asked Robin.

"Yes, weird things happen to Roger every May twenty-ninth," said Paul.

"Not every May twenty-ninth," I replied, "and weird things happen on other days of the year. Like the blizzard in Tierra del Fuego."

"Oh, yeah, Paul told me about that," said Robin. "You guys got lost in a blizzard and ended up holding hands and praying or some such thing."

" 'Holding feet' was more like it," said Paul. "At least I held Jack's feet—in my down jacket because they were frozen."

"Sounds serious. So you prayed and the storm stopped?" asked Robin.

"Peter called it a concentration, some kind of Tibetan Buddhist thing," said Paul. "I guess some people would call it creative visualization, but it was more than that."

"I've been looking for a scientific explanation," I said.

"Aha, another godless scientist," said Robin. "How could there be a scientific explanation for such a thing?"

"I'm not godless, but also I believe we are here to increase knowledge," I said. "At the smallest imaginable scale of the universe there is a world of quantum waves that we cannot see."

"So you think our minds are linked to everything else through the quantum waves." Robin took a sip of wine.

"Yes," I replied.

"That would mean everything in the world is connected; obviously it isn't. How can that be?"

"I'm not sure, but I think it's like in a laser. Individual light waves join together to make a single powerful wave. In Tierra del Fuego the blizzard abated at precisely the time we thought we were finished, at the point when we couldn't go on, when we had given up hope. Somehow that allowed our thoughts to link to the storms."

"So your thoughts moved the clouds?" asked Robin.

"Either that, or we shifted to another reality where the clouds didn't exist. Probably both explanations are the same."

"Another reality, you mean a parallel universe?"

"Yes, we chose a reality with a different ending," I replied. "I think all possible futures exist simultaneously in the present. That is the only way precognition can work."

"Are we talking about another May twenty-ninth experience here?" Robin drained her cup.

"More like a couple of them. Guruji in India predicted the date of Brendan's birth. He's a May twenty-ninth experience." Maggi laid Brendan's stick on the ground and picked up her wine cup.

"Perhaps we should leave before we become May twenty-ninth experiences." Paul laughed.

"Aha, more victims. It helps if you're drunk." I waddled toward Paul like Dr. Frankenstein's assistant, Igor, and filled his cup. "Drink! Thinking bad! Wine good!"

"Paul never gets drunk," said Robin.

"Perhaps not in this universe," I mused.

"No, I'm interested. Brendan's birth was a May twenty-ninth experience?" asked Robin as I filled her cup. "What's that have to do with parallel universes?"

"Apparently some part of us chooses our own futures from an infinity of possible universes." I filled Maggi's cup. "We create time by experiencing a sequence of futures."

"Infinite futures, you realize how crazy that sounds?" declared Robin.

"Yeah, I know; I hardly believe it myself. Drink your wine." It is the only logical conclusion.

"Infinite futures." Paul jumped up and spun around, looking up at the moon through the trees.

"Only because we haven't mapped it into our mental landscapes yet," I responded.

"You're right about that. Why aren't we aware of these other futures?" said Paul.

"We are barely aware of anything in this universe," I replied. "Ninety percent of it is dark matter that we have never seen. Furthermore, we never notice anything that takes place faster than a sixteenth of a second—the universe reached over half its present size in less than a sixteenth of a second. Image all the universes that could explode and collapse and we'd never notice them."

"Subliminal universes? Why's that?" Paul sat back down and took a sip of wine.

"It takes our brains that long to process the information from our senses, to present our consciousness with a simulation of reality." I picked up Brendan's stick and waved it in the fire.

"Our brains simulate reality?" Robin asked.

"We are only conscious of our brains' interpretation of reality, an illusion."

"So you are my illusion, and I'm yours?" asked Robin.

"Or maybe allusions," I replied.

"Sound's like you are both deluded," Paul responded.

"Thanks, very funny," I said, laughing. "We each access similar sensations, but our brains can give them different spins."

"We both see the same moon," said Paul.

"Sure, because our brains are similar, but I assure you crabs and frogs experience different moons from us."

"If we can't count on the reality of the moon, what can we count on?" asked Robin.

"The moon you see now is not the same one you saw three hours ago," I replied.

"The moon is a simulation constructed by my brain?" she asked.

"Take the size. Is the moon the same size now as it was on the horizon?" I pointed to the moon that had just risen into the open sky above the treetops.

"It's smaller now," said Robin.

"It's actually the same size, as you can verify by measuring it with a ruler held at arm's length." I held my thumb to the moon and sighted across it.

"It sure looks smaller now," said Paul.

"That's because our brain automatically adjusts the size of things to the distance above the horizon, to keep things in perspective," I explained.

"Next I suppose you'll tell us that sunsets are illusions." Robin looked over her shoulder for a hint of redness in the sky.

"In a sense. The color is an illusion, a phantasm of perception. Our brains have a built-in white balance. The setting sun would appear much redder, but our brain automatically constructs a green filter to block out the extreme redness. If you look away from the sun quickly, you can catch a glimpse of the green filter, the same size and shape as the sun, a kind of mask."

"The famous green flash!" said Robin.

"Perhaps." I took a sip of wine.

"So you are saying that there are Big Bangs, universes being born all around us, and that our brains ignore them?" Paul wanted to know.

"Yes. But, if you think about it," I said, "common sense requires that there must be more than one universe."

"Why's that?" asked Robin.

"Because the universe as we know and love it is far too perfect," I replied.

"Perfect? It seems far from perfect. Look at all the empty space out there." Robin waved vaguely in the direction of Cassiopeia.

"We only learned how perfect the earth is by learning how imperfect our neighboring planets Venus and Mars are. We are living on the only planet with liquid water. On Venus water is steam, on Mars ice." I drained my cup and set it down on a boulder.

"And liquid water is necessary for life," said Robin.

"At least, life as we know it." Paul finished his wine and let the cup dangle from his finger.

"Everybody knows that Venus is perfect," challenged Robin.

"From far away it is. But up close, it's too hot. The atmosphere contains sulfuric acid." I poked Brendan's stick into the fire.

"Talk about acid rain!" exclaimed Robin.

"I get it," said Paul. "Venus is too hot, and Mars is too cold."

"And Earth is just right—the Goldilocks planet," said Maggi.

"Not only in temperature. The atmosphere on Earth is just the right density and composition." I pulled Brendan's stick out of the fire, stood, and waved it through the air. It burst into flames. "See, oxygen! That wouldn't happen on Venus or Mars. The list goes on and on. I feel like Ranger Rick giving a fireside talk."

I sketched a letter *B* on the blackboard of the night. "Class, what letter is that?"

"*B*, Ranger Rick," the three of them replied in unison.

"Exactly. Your brain smoothed out the individual points made by a single stick into a continuous letter."

"Yes, and now I see green *B*s flashing out of the woods all around me," joked Paul.

"I estimate you are only one sip of wine away from seeing universes all around you." I filled his cup as he held it out.

Paul took a sip. "You're right, they *B* everywhere."

"Very funny. I always say—a bad pun, better than none." Nonetheless, I couldn't help smiling to myself.

"You're doing good, Ranger Rick," Paul said.

"That's what I do," I responded

"What do you do?" Paul asked

"I do campfire talks. A lifelong dream."

"I get the idea," said Robin. "We live on Earth because it is the only place we can live."

"If you think about it, the perfection of the universe is even more amazing than the perfection of the earth." I swept my pointer across the sky.

"I'm sure you'll fill us in," said Paul.

"Tell us, Ranger Rick," said Paul.

"There is the question of the chemical elements, the atoms in our bodies."

"Weren't they produced in the Big Bang?" asked Maggi.

"Hydrogen was, over ten billion years ago."

"No wonder I feel so tired," murmured Maggi.

"You're a mother; you're supposed to feel tired." I pointed to the tent with my stick. "But all the other elements, with the exception of some helium, were produced in stars, mainly exploding stars, by combining lighter elements to make heavier ones. If the force that holds the nucleus of atoms together were much less, stars could not produce the heavier elements. The universe would be only vague clouds of hydrogen. On the other hand, if the nuclear force were much stronger, only the heavier elements would have formed during the Big Bang. We'd have no hydrogen, no water, no marshmallows."

"That hurts," said Robin

"Your point?" Paul feigned tedium. "If there is one."

"Sorry. The point is that we have marshmallows; the nuclear force is just right."

"Another Goldilocks effect," said Maggi.

"There could be all kinds of universes, ones that collapse and burn out immediately, ones where gravity is so weak and matter so scarce that they never pull together into stars or galaxies," I explained. "The fact we have galaxies in our universe, the fact we can look up and see stars, the fact that the universe has been steadily expanding at just the right rate is extraordinary."

"What's that got to do with parallel universes?" asked Robin.

"In our universe, the forces of gravity, electromagnetism, and the strong nuclear force are finely tuned, to the one-hundredth part of a decimal. Our universe is as perfect as it can be."

"So God made the universe perfect, just for us," said Robin.

"That's one accepted idea," I replied.

"And the scientists' idea?" asked Paul.

"Until now scientists have assumed that the perfection of the universe was an accident."

"Until now?" pressed Paul.

"Now there is the multiuniverses idea."

"Which is?"

"That all other universes do exist—they come and go, explode and implode, or drift away, nondescript—but by natural selection and billions of years of evolution, we live in one that is just right for us."

"I still find it hard to accept." Robin shrugged and looked up at the stars.

"Is it any stranger than the idea that God created heaven and earth in six days? I mean how did God decide when and where to create heaven and earth. Why six days?"

"The Bible is not meant to be taken literally," said Robin.

"You are used to the biblical story, so it does not seem fantastic. The multiuniverse story seems strange because this is the first time you heard of it."

"It's certainly no stranger than the idea that the universe is a total accident, an unexplained coincidence," added Paul.

"Roger doesn't believe in coincidences," said Maggi.

"I believe in some of them," I explained, "but I don't believe the universe is a coincidence. And I don't believe that the fact that the blizzard stopped just after we made our concentration was an accident."

"Neither do I," said Paul.

"Why can't you just call it a simple miracle?" asked Robin. "You make it sound like we can just change universes at will."

"Supposedly, every time we make an observation, a new universe erupts with a carbon copy of us in it. We are only conscious of one universe at a time."

"Why aren't we aware of all the copies of us in all the other universes?" Robin drained her cup and stood up to turn her backside to the warmth of the fire.

"It would drive us crazy. We couldn't handle all the possibilities. I think that is why I was so freaked out during the accident in Morocco."

"Another May twenty-ninth experience?" asked Robin.

"It's a long story," I replied.

"Our egos have invested so much in consciousness's worldview that it protects its investment at all costs, even if it's wrong—especially if it's wrong."

"What's the purpose of changing worldviews?" asked Robin.

"Perhaps we can learn to change universes without going insane."

"So you think that there is some part of us connected to the world of many universes?" Paul stood up next to Robin and rubbed her back.

"Yes, I think thoughts are quantum waves connected to the quantum foam from which the universes erupt. We cannot be conscious of the unconscious. Consciousness is overrated," I said.

"Perhaps the May twenty-ninth experiences were just a series of lucky happenings," said Robin.

"Luck? I don't think so. At each point of teaching, extraordinary people came forward, Chombi and Guruji, for example. Not to mention Paul and Maggi. I know there is a part of each of us that watches the whole thing, a cameraman, recording these events, the computer programmer who directs the attention of our consciousness." I spread the fire with the stick. It burst into flames.

"Like Guruji said, we have to let go of our attachments," said Maggi.

"Yes, but the desire to let go is an attachment," I said.

"Let's get out of here before we become May twenty-ninth victims." Robin stretched, and craned her neck to the heavens.

"We have to leave at daybreak. It was good seeing you again." Paul hugged Maggi, then me.

"Abrazos, amigo!" I said.

"Let us know if you survive tomorrow." Robin hugged us, put her arm around Paul, and together they turned to their van.

"See you guys." I kicked dirt over the glowing embers. A chill went up my spine as the last coal faded. The cold beam of my flashlight illuminated plumes of dry smoke puffing through the shroud of dirt, dust, and ashes, like in the valley of ten thousand smokes.

"God, I'm cold." A brisk wind flapped at my shirt. Maggi was already through the flaps of the tent, and I followed immediately. We crawled into the sleeping bag, and we curled up in each other's arms. I heard the door of the van door slide shut. A gentle wind picked up and murmured in the fir boughs.

"This is the universe for me. What about you, Goldilocks?" I said as I dropped off.

"Yeah, it is," Maggi whispered.

I dreamed of a great flood of blinding energy held back, dammed up by the fir trees around our camp. I floated up into the trees, now sparkling with blue-and-purple auras. Streaks of light raced down the trunks from the great white light into the earth, where it shot back up, held me aloft, raced through me, and, leaving through my head, surged back up into the tree branches.

The next morning, the alpine snowfields, melting in retreat toward the summit, uncovered matted-down prairie grass on their fringes. Lower down, where the grass had already sprung up, egg-yellow snow lilies twisted toward the sun. Maggi and I and our three sons followed an elk trail upward along the fringe of blue spruce trees that bordered the prairie. We stopped and looked at the patchwork of clear-cuts quilted over the mountains of the Coast Ranges.

"What a depressing mess." I held my camera to my eye and snapped pictures of the clear-cuts.

"Remember, they're just illusions," said Maggi.

"I wish."

We followed the trail into the shaded canopy of evergreens and rested on a large nurse log.

"Lemon Head break; who's a Lemon Head?" I held out a bag of yellow lemon-shaped candies.

"Me! Me! Me!" the boys replied in a chorus.

Chickadees twitted in the branches overhead.

"Hey, my chickadees are here; they want Lemon Heads too," I said.

"Really?" asked Ian.

"Just kidding; they eat the seeds from spruce cones." I laid my camera down on the old log. "That's why they always stay in the tops of the trees."

Brendan picked up my camera. "I'm going to take pictures of the chickadees."

"Good luck; they won't show up from here." I pointed to the high branches.

Brendan snapped a picture and set the camera down.

"I want to take picture," said Kieran, leaving Maggi's side and reaching for the camera. It dropped out of his hand and rolled in the needles of the forest floor.

"Damn it!" I screamed, and slapped Kieran's hand. "Be careful; that's not a toy!"

Kieran looked at me. Conflicting emotions of fear, confusion, and shame pulled his face in different directions. The flood of emotions built up, and he broke out bawling.

"Roger," Maggi admonished me gently. "You did let Brendan use the camera."

"He's older; he doesn't drop it on the ground," I said. "Ohh, you're right. Kieran, I'm sorry; come here, buddy." I hugged him and stifled his sobs in my chest. Spasms of sighs and tears tore at the body ravaged by the injustice. He tried to stop crying, but couldn't.

A chickadee flitted through the forest and landed on a thin branch near my face. His brown eyes held galaxies of compassion.

"Look, Kieran," I said. "Look, here's a chickadee."

He turned, reached for it, and immediately stopped sobbing. The chickadee flitted up to avoid Kieran's reach and landed on my head.

"Quick, Kieran, take his picture. Chickadees never land on people's heads." As if posing, the chickadee sat in my hair while Kieran picked up the camera and snapped a picture.

"Look, the chickadee thinks Daddy's hair is a nest," said Ian.

"Well, there's your May twenty-ninth experience," said Maggi.

"I hope so. I've had enough near-death experiences for one lifetime." The chickadee took off and disappeared in the canopy of spruce boughs.

I went on to raise my family on the Oregon coast while doing research and teaching at Oregon State University's marine science center, in Newport. Not until my youngest boy entered college did I finish my analysis of the May 29 experiences.

CHAPTER TWELVE

RELATIVITY'S RAINBOW

In 1982, while living in Oregon, I received word that my father was in the hospital. After delivering a paper at a meeting of the American Geophysical Union in Toronto, I flew to Boston with Brendan, eight years old at the time.

I peeked into the hospital room, hesitantly, holding Brendan's hand. My father slept in the fetal position between sheets stretched as tight as restraining straps. A pale sliver of his upper back showed through the green hospital smock, an IV was taped to his arm, and a plastic oxygen cannula hung in his nose. A TV on a corner shelf near the ceiling was tuned to *Let's Make a Deal* with the sound turned off.

He won't be happy to see me, I thought to myself. I sat down timidly in a chair next to his bed and lifted Brendan onto my lap. Brendan's blond hair, grown in long, was cut like Prince Valiant's.

My father rolled over onto his back, ran his hand over his eyes and sparse white hair, opened his eyes, and stared at the far wall corner. Shaking, he grabbed the steel bed rail, pulled himself up, rolled his head, and glanced at me. His hazel eyes lit up with the smile of morphine. He seemed genuinely pleased to see me.

"Oh, Roger, how are you?" He lay back, relaxed, and draped his hand over the rail.

"Fine," I said. "And you?"

His eyes lit up, and he tried to push himself up again. "I saw him up there," he whispered in a weak voice.

"Up where? You mean on TV, Monty Hall?"

"In the other corner, there." He lifted his arm slowly with plastic tubes hanging down and pointed to the corner where the white ceiling intersected the two white walls. "I saw him up there."

"Who? Who did you see up there?" Brendan and I stared into the corner but couldn't see anything.

"Pop. Did you see him?" My father spoke slowly.

He must be hallucinating, I thought to myself.

"Pop?" I asked.

"Yeah, Pop," he replied.

"Grandpa?" I assured him. "Yeah, I think it was him. I'm sure it was him." I stood Brendan up on my lap. My father noticed him for the first time, smiled, and reached stiffly toward him. "Who is this handsome boy?"

"This is Brendan," I replied. "Say hello to your grandfather, Brendan."

"Hello, Pop Pop." Brendan sat down and leaned shyly back against me. I put my arms around him, squeezed him, and blew on his neck; he squirmed and smiled meekly. He wore an orange vest I had bought for him in India, embroidered with peacocks and inset with tiny mirrors,

My father, hand shaking, groped for Brendan's, and grasped it firmly. Brendan turned and tugged on his hand. I whispered in his ear, "It's okay. He's your Pop Pop."

My father held on to Brendan's hand for a few moments and let it go. I took my father's hand and held it. I thought that he would pull it away and was surprised when he didn't. To the contrary, he looked up at me lovingly.

A nurse in a tight-fitting white uniform walked crisply into the room, checked the morphine drip, and patted my father on the chest. "How are you doing, Mr. Hart?"

"Fine." He beamed lovingly at her and pushed up to watch her walk out of the room. "You never lose that."

"Lose what?"

"Sex, interest in sex." He gave me a mischievous grin.

Oh, I thought to myself, like me. I never imagined my father as a sexual being, let alone in a state of sexual arousal at the age of eighty-two. Must be the morphine drip. He stared at the TV.

Monty Hall wandered through the audience, microphone in hand, while costumed members of the audience—fluffy chickens, shaggy dogs, hairy leprechauns—jumped up and down and waved their arms to get his attention.

My father's eyes moved from the TV to the corner where he had seen his father. The focus left his eyes, he closed them in rest, and after a time they came back to me, sparkling again.

He looked at me earnestly. "Do you believe in hell?"

"Hell? It's all in the mind," I replied. "Invented by religions to prevent religious experiences." I'd come to the realization that to follow the dogma of either devil or angel is an emotional crutch taken up at the expense of tremendous personal power and knowledge.

"That's what I believe," he replied. "What about Satan? You know I never went to church. I was never much on religion."

"Satan is a projection of guilt," I replied. "And guilt is a useless emotion."

"Yeah, right. Oh, guilt." He pulled his hand away and curled into the fetal position. I thought he had fallen asleep, but he turned his head to me. "About your mother. I'm so sorry."

"Sorry?" I sat Brendan back on my knee and bounced him nervously.

"I locked the doors, but she started climbing out the windows." He reached out with his hand trembling.

"Climbing out windows. What for?" I asked.

"I couldn't cope with her anymore," he replied. "She'd run down the street in her nightgown. At first they said they were just going to keep her for tests."

"You mean, test her for premature senility?"

"Like her father. There was nothing I could do. They moved her to Danvers. Within a month she died, in the same place as her father." A tear ran down his cheek.

I took his hand between my two. "Dad, you did what you had to do. It wasn't your fault."

"I just didn't know what to do." He lifted his trembling hand toward me and looked at me beseechingly.

"It's not your fault she died." I stroked his hand. "No blame. No guilt."

"No blame?" He let his hand go limp.

"No blame." I let his hand slide down to his side.

He closed his eyes and patted the hair on his head, his face white as chalk, his cheeks sunken in around his jaw, a tangle of wrinkles stalking his eyelids. Without opening them, he mumbled, "You know, we thought you shouldn't have gone with Sayre."

"Yes, I know," I replied. "Actually, you were right; I should have listened to you. I'm sorry; I just had to figure things out on my own."

"Sure you did. I understand, glad you did." He opened his eyes and looked at me. "Sometimes I wish I had more time to figure things out. To do what I wanted to do."

"What do you mean?" I asked.

"I wish I spent more time in science. Take Einstein, he had it made."

"What do you mean?"

"All those crazy ideas, quantum mechanics and relativity, accepted in his lifetime," he continued. "He was a genius. He got to do what he wanted. I never got to do anything I wanted. Did I ever tell you that I invented radar?"

"You invented radar?"

"At Raytheon, the vacuum triode that could detect phase shift in radio signals. Of course, some guy in D.C. beat us to the patent."

"But you discovered it first?"

"I'm not sure first—independent. I discovered it independently."

"That's fantastic," I said. "That's the important thing. You thought of it on your own. How did you feel?"

"The greatest feeling I had in my life—except the feeling when Ruth brought you kids home from the hospital."

He smiled weakly, and then his face lost all expression. "You know something?"

"No, what?

"I think, I think I'm going to die. I'm so sorry," he whispered.

"Sorry?" I asked. "For what?"

"Sorry, that I'm dying. Sorry, that I'm failing you."

"You have nothing to be sorry for, and you never failed me. You made everything possible. Like Mom always said, us kids have your brains."

"I'm, I'm a little scared." His eyes flared up.

"Don't be scared. Believe me, it's all downhill from here. Just relax. Let go."

"Oh, yeah." He smiled mischievously, winked, and closed his eyes. "Easy for you to say."

He rolled his head to the side and closed his eyes, and the color drained from his cheeks. He let his hand fall from mine. He looked helpless, innocent, like a baby asleep.

"Is Pop-Pop dead?" Brendan turned away from my father and looked up at me in alarm.

"No, he's just resting."

Monty Hall was pointing to three doors on the stage while a dog like Pluto jumped up and down and clapped his hands. I bent over and kissed my father for the first time in over thirty years, kissed him at will because he did not notice.

The next morning, as I brushed my hair in the bathroom of my father's house, a puffy red face staring through the window startled me. Dick McDermott. I hadn't seen him since the day he threw me off his lawn fifteen years ago.

I waved to him. "Go around to the front door," I shouted through the window, and made circling motions with my arm.

The door chime went off when I opened the door. I shook hands with Dick on the front step and introduced him to Brendan. "Hi, Dick, meet my son Brendan."

Dick bent over, held out his hand, and Brendan shook it.

"You're a good boy; I can see that." Then he stood up and looked at me. Tears welled in his eyes.

"How's your father?" he asked.

"Ready to move on," I replied.

"Oh, dear, everybody is passing," he said, eyes lined with red. "My wife, Edith, died six months ago."

"I'm sorry." I put my arm over his shoulder.

"I can't believe that she's gone, after thirty years." He put his arm over my shoulder.

"Do you believe in heaven?" he asked. We turned and walked toward the driveway.

"Of course," I said.

"I haven't been that much of a churchgoer. I've been such an asshole."

"Me too. It doesn't matter."

"Oh, God, it's so sad." He wiped the tears from his eyes and buried his head on my shoulder.

A pair of Baltimore orioles flitted through the branches of the black oaks shading my father's lawn. A dark sky, morose with humidity, hung in the tree branches. A sob surged through Dick's body; he straightened up, patted me on the chest, and said, "I've kept the lawns mowed for your dad." He walked me to the flagpole lawn. "See." He swept his arm across the trim grass.

"Thanks," I said.

"Thank you." He turned toward his house, head hung low.

"It was nice talking to you," I replied.

"Yeah, see you later."

A clap of thunder shook the heavens, and a cold breeze swept the grass, rattling the ropes on the flagpole.

I turned to Brendan. "We better take down your grandpa's flag so it won't get wet."

I showed Brendan how to lower the flag on the pulley. A few drops of rain landed on us as I taught him how to fold the flag the way I had learned in Boy Scouts. As we crossed the driveway, a heavy warm rain drenched the asphalt and danced in sheets to the manhole. There it backed up and pushed and shoved to get through the grating, flowing into the underground labyrinth of surge drains. How many times, I thought to myself, had the ocean waters swept the land and washed the refuse to the bottom of the sea? At least ten million times since the formation of the earth.

Brendan marched up the flagstone walk carrying the flag in his outstretched arms. I lifted him up so he could place it on the front hall closet shelf beside my father's Navy captain's hat.

"Let's go watch TV," I said.

Brendan sat on the floor of the upstairs sitting room watching TV while I sifted through duffel bags of cold-weather gear from the Antarctic, Tierra del Fuego, and the Himalayas in the closet of my old bedroom. I dragged the remnants of my Mount Everest gear to the room where Brendan was watching *Sesame Street*.

"Yum, me love cookies," the Cookie Monster was saying.

"Look, here's the parka I wore on Mount Everest," I said to Brendan. He looked up at me, then turned back to the TV.

"Here, Brendan. Stand up. Try it on." The puffy blue parka hung to his ankles. He looked like the Michelin tire man.

"Cookies! Cookies! Me want Cookies," Cookie Monster said. Brendan slipped off the parka and looked back at the TV.

"Oh, look, the boots Chombi gave me. Here, try these on; they're from Tibet!" I said.

He sat down, pulled them on, all the while looking at TV. They came up to his thighs. I scraped around in the bottom of the duffel bag and pulled out a plastic bag with hairs in it.

"Look at this," I said to Brendan, "yeti hairs."

"What's a yeti?" He asked without taking his eyes off the TV.

"An animal, an ape that's like a man," I replied.

"You mean a monster, like Cookie Monster?" he asked.

"Yes, the yeti is just like Cookie Monster except he can't get a job on *Sesame Street;* his English is bad," I replied.

"He can't talk English?"

"No, he speaks Urdu or something." I stuffed everything back into the duffel bag.

I pulled the family picture album out from under the TV stand and leafed through the pages. Faded brown photos depicted scenes of family history: my father sitting in the cockpit of a biplane that looked like a Wright brothers original, my father at the controls of the first amateur radio station at MIT, the whole family together around a picnic table at a White Lake campsite, me and my mother pretending to be asleep on granite beds on top of Mount Chocorua.

A separate scrapbook contained newspaper clippings. A 1960 article in the *Explorer's Club Journal* detailed the expeditions to Antarctica I was on with Professor Nichols. A clipping from the *Christian Science Monitor* described the glaciers named after us in the Wright Dry Valley of Antarctica.

"Look, Brendan, here's Daddy's glacier!" He glanced at me and turned back to the TV.

"Hey, Brendan, do you know what a glacier is?" I asked.

There was no response. I found a clipping from the front page of the *Boston Globe,* June 22, 1962. At the top was a photo of Woody, me, and Norman with Sherpas Aila and Pemba taken in Katmandu. The headline read, "Climbers Found Safe, Weak," illustrated with an artist's drawing of a wooden lean-to where we were supposedly found beneath a mountain peak. The caption read: "Spotted by American in Plane, Huddled in Thatched Hut in Storm-lashed Nepal." I doubted if there was a shred of truth in the account. The newspaper not only made up the story that we were lost but also the circumstances of our "discovery." In fact, we walked to Chombi's house in Khumjung and radioed for a helicopter. Norman Dyrenfurth, the leader of the 1963 "First American Everest Expedition," flew with the chopper pilot to show him the way to Khumjung. He was widely credited with our rescue. My God, "a thatched-roof house in storm-lashed Nepal." How did the reporters imagine the roof stayed on?

I read through a clipping from the June 23, 1962, issue of the *Boston Record American* that showed my mother and father talking on the telephone, posed as if receiving news of our safe return in Nepal. They were both smiling and looking good.

Brendan wasn't interested in my mementos or photographs. Okay, let's change the subject. I tickled him, picked him up, and swung him around like an airplane, one hand under his chest and the other holding his arm stretched out like a wing.

"Brendan," I said. "*B* is for *dive bomber,* Brendan." I swooped him along close to the floor. His hair hung down and he squealed with delight.

"Now, you're flying over Mount Everest." I held him high and spun him around. "Now you're flying south over the ocean. Oh, my God, it's getting cold. Look at all the ice; that's Antarctica. Oh, my God, you're going to freeze your nuts off. Hang on to your nuts; it's getting cold."

"What's my 'nuts'?"

"Forget it. You're okay. We're flying down the Wright Dry Valley. Look! Look, there's Daddy's glacier."

"Who gave it to you?"

"Oh, we're too low." I made noises like an engine pulling out of a tailspin. "*Brrrrum*—the United States Board on Geographic Names gave it to me—faster *Brrrrrum.*" I spun around faster and faster, sweeping him along at arm's length.

He shrieked, "Faster! Faster!"

"*Phew,* we made it, we're safe." I collapsed on the floor with Brendan on top of me.

"More! More! Let's fly some more, Dad."

"No, sorry; engine's out of gas. Come on let's go to the woods. I'll show you my woods. Did I ever tell you, I'm king of these woods?"

"Your woods?" he asked.

"That's right, I'm the king."

• • •

We stood on the granite ledge behind Dick McDermott's house and looked over the green canopy of oak trees to the sparkling water of Walden Pond. The rain had washed the sky clean. Buildings of Boston traced the skyline to the south.

I picked up a stick, stamped it on the earth like a scepter, and proclaimed, "The king and his fearless knight have returned! You be the knight." I turned to Brendan.

"I am the knight." Brendan picked up a stick.

"Yes. We've come back to . . . ah . . . to drive out all the monsters." I stamped my scepter again.

"To drive out monsters." Brendan parried the air with his stick.

"Watch out, monsters, here we come!" We ran down a ledge, leaping over boulders. As we crossed the main dirt road around Walden Pond, several black trash bags stuck out of the bushes. "Look at that trash the monsters left." I whacked the bags with my scepter. "Begone, trash!"

"Yeah, get out of here, you mean old trash." Brendan waved his sword at it.

"Look at that beat-up old couch the monsters left," I said. Brendan stabbed it with his sword.

We ran along the path that skirted the shore of the pond and climbed out on a rock overhanging the water. The shore stank of slime, and the trunks of drowned trees jutted above the surface.

I picked up a stone and threw it in the pond. "Get out of here, mean eels."

"Yeah, mean eels. What's an eel?" asked Brendan.

"A fat snake with fish fins," I said.

"Yuck, get out of here, mean eels." Brendan launched a stone toward the water, but it splattered into mud on the shore.

"Get out of here, acid rain." I threw stone after stone into the lake.

"Yeah, what?"

"Acid rain; it's mean yellow smoke that falls out of the sky and hurts the trees. See how sad they look."

"Yeah, get out of here too!" Brendan threw a stone in the pond.

"Out, pond scum!"

"Yeah, pond scum," Brendan repeated.

"Bug off, snappers!"

"What's a snapper?"

"Turtles that can bite your foot off."

"Yeah, snappers." He looked up at me with horror on his face.

"Okay, we've purified the pond; let's go to the castle on the hill." We took off running along the path. "Hurry! We pissed off the monsters. They're after us," I yelled over my shoulder.

"I hear them. Run! Run!" shouted Brendan.

For a moment, I imagined I heard twigs snapping and the sound of running in the woods. "Yeah, run! Run! Let's run from the monsters!"

We stumbled over logs, twigs slapped us in the face, a rush of adrenaline surged up my spine, and the hair on my neck went up as if wolves streaked through the trees on either side. We were alone, anxious, in a world prey to ravenous predators. Our shoes scraped on the slick granite as we scrambled up the hill, sweating and panting. We reached the top and made our stand.

"Look, an old airplane!" I said. The dull aluminum frame of the Piper Aztec—paint peeling, windows smashed—sat askew in the bushes, one wing tilted up on the branch of a black oak.

"That's not an airplane," said Brendan. "That's a castle that fell from the sky. Quick, get in. We'll be safe."

We climbed into the two moldy seats with springs pushing stuffing through the torn fabric. A sudden feeling of loneliness and despair went through me. My father would soon be better off, in a better place, but I felt lonely, helpless, and sad. On the one hand, my father's imminent death released me from his accusations of not focusing, of not being serious. I was no longer a "jack-of-all-trades, master-of-none." I was no longer "a boy sent to do a man's job." But at the same time, I felt alone, left behind, abandoned. I began to sob. I turned away from Brendan and held my hands over my eyes.

"What's the matter, Dad; are you afraid?" asked Brendan.

I took a deep breath. "No, not at all."

"I'm scared," he replied.

"There's nothing to be scared of." I wiped my eyes and looked at him. "We're just playing."

"What about the monsters in the woods?" he asked.

"They're just in our heads. I'll get rid of them," I said.

"A purge! A purge on all of you!" I yelled out. "Without fear, without greed, I face you monsters. Now leave these woods."

"Are they gone?" he asked.

"Don't think about them, and they'll go away," I replied.

"Okay, I'm not thinking about them anymore."

"See, they're all gone," I said. "It's safe to go home and watch *Sesame Street.*"

"All right."

I took Brendan by the hand, and we walked through the woods. I took comfort in the indestructible forms of nature inscribed everywhere—millions of identical oak leaves, thousands of wintergreen plants with the same aromatic molecules producing the same smell, every blueberry bush the same height. In those woods it was obvious that the mind of God maintains the templates of symmetry that defy chaos in every hemlock needle.

At some level we are at play in the quantum field, taking one form, now another, spontaneously, joyfully increasing the awareness of the beauty in the universe just by looking at it. I inhaled the air perfumed with perfect sweet fern. I held Brendan's hand, and skipping along the blueberry path, we sang "Can You Tell Me How to Get to Sesame Street?"

The impact of my father's death was deeper than I could have imagined; I didn't realize how much he had helped define my life until he was gone. I was thankful that we had reconciled our differences during the visit in the hospital. Strangely, although I was certain he lived on in nonmaterial form, his death took an enormous emotional toll on me. I knew my own thoughts continued after my consciousness left my body, but I could not fully grasp the comfort of that knowledge.

I pondered the question of whether or not it was possible for science to prove there is a "life" after death. The separation of mind and body during near-death experiences might be proved by establishing a coordinated study with routine monitoring equipment set up in trauma centers, emergency rooms, and operating rooms of a significantly large number of hospitals. The rooms could be equipped with specifically designed images set up—out of the line of sight of the operating table—on the top side of lights, fans, and pieces of equipment. Oral histories could be recorded of patients who had recovered from flat-lined conditions to determine if they had a recollection of the hidden images. Care, planning, and considerable work (say, interviewing a thousand patients a year for five years) could provide evidence of separation of mind and body—an indication that some form of consciousness survives death.

Before a rigorous investigation of the survivorship of consciousness can be made, it is necessary to better define consciousness. My near-death experiences convinced me that there is more than one kind of consciousness. At least one kind is closely associated with the functioning brain, but there is also at least one kind that exists independent of the brain.

Consciousness that generates behavior is different from consciousness that facilitates the experience of experiencing. Only by identifying and naming its component states will it be possible to discriminate conscious states generated by the brain from conscious states affiliated with the nonmaterial mind

Consciousness is like the light from a flashlight; we are aware of only what the light illuminates. All else is in the realm of unconsciousness. We cannot be conscious of the unconscious, or can we?

If we sweep a flashlight around a dark room, we are aware of only what the flashlight illuminates. Yet we can direct the beam of the flashlight from one object to another. Our mind interprets and records each object in our unconscious memory. We recognize these objects because of previous experience with them or similar objects. That's a chair. That's a round table, three feet tall. And so on.

Eventually we can cover the entire room with the flashlight, recording the design of the wallpaper, the types of furniture, the patterns of the rug, the number of ceiling fixtures; in our mind we can assemble individual views into an overview of the entire room. We can sit down with a pen and paper and draw a map of the room by mosaicking together the individual frames recovered from the unconscious. We can then look at the completed map and gain a conscious impression of the entire room that resembles the room if we turned on the overhead lights and observed it. We have expanded our consciousness by use of time and memories stored in the unconscious mind. We can preserve the map of the room as a cultural record.

The contents of the complete map of the room constructed by our consciousness depend on where we shine the flashlight, how we interpret what it illuminates, and how the scene is reconstructed from memories. There could be a different type of consciousness corresponding to each function: that directing the flashlight of consciousness; that deciding what scenes to record; that deciding which scenes to retrieve from memory. Shouldn't each type of consciousness be given a different name?

My May 29 experiences led me to infer that our brains construct a simulation of reality. Our conscious interpretation of the world is based partially on past experience. Even members of our own species, with similar brains, will not see the picture of the room the same way we do. When members of indigenous tribes in New Guinea were shown pictures drawn on paper, they saw only the paper, not the picture. When African Bushmen, those not accustomed to expansive vistas, were guided out of the forest for the first time, they saw large elephants in the distance but thought they were small elephants close-by. The Bushmen would not relate to perspective in a drawing. Is their conscious experience less real than that of others who perceive differently?

There is a famous story of Picasso riding on a train. He sits down next to a man and introduces himself. The man recognizes Picasso as the creator of weird paintings of women with both eyes on one side of the head or perhaps a leg emerging from the neck.

"Why don't you paint things as they really are?" asks the man.

"What do you mean?" asks Picasso.

The man pulls out a photograph of his wife. "Why don't you paint women like this?" He shows the photograph to Picasso.

"She's rather flat and small, don't you think?" replies Picasso.

It was necessary to come up with a plausible mechanism of how our brains simulate reality. A plausible mechanism is inherent in the way the brain works. It works as a filter by discarding information; eleven million bits of information are discarded every second.

Imagine that a source of white noise, such as the traffic at a busy intersection or a lilting cataract on a mountain river, is analogous to the information that enters the brain. All notes of the musical scale are embedded in the white noise, the high note of a flute as well as the deep boom of a bass guitar. It is theoretically possible to devise a system of sound filters that block out all but a few notes. By choosing the right sequence and timing of filters, it would be possible to reproduce Beethoven's Ninth Symphony or imitate Jimi Hendrix playing "The Star-Spangled Banner." All other possible notes, all possible songs, still exist in the white noise, but they have been blocked from consciousness by our filtering system and our decision to focus on Beethoven or Hendrix. Our brains are like this filtering system. We can make choices, but they are limited. We can choose Hendrix or Beethoven, but not both.

The brain allows only one aspect of quantum waves to enter consciousness; all other aspects continue to exist but are filtered out. The brain might even filter out one of a pair of destructively interfering quantum waves, allowing the other to manifest as matter seemingly out of nothing. We experience one member of a complementary pair of aspects—momentum or position, energy or time—but not both. Momentum manifests when position is filtered out. Time manifests when energy is filtered out. Superluminary speeds or branching universes are not needed to explain the collapse of the wave function in the filtering process. All aspects of the complementary wave function continue to exist, one in consciousness and the rest out of consciousness.

Our brain presents us with a simulation of reality, and we don't seem to have control over what interpretation it presents to us. However, we can say no to, reject, any given simulation.

The second eureka moment associated with my analysis of the May 29 events came while I was walking along a creek in the rain forest near my home in Seal Rock. The amber crescendos of Swainson's thrushes arose from the alder groves around me. I was thinking about a question that had perplexed me for some time. If the reality we experience is an illusion, a simulation, created by our brains, how is it that we share common experiences with other individuals? The staccato cry of a pileated woodpecker arched over me as the red-crested bird darted from one Sitka spruce to the next. Footprints of elk embedded in a muddy bank trailed off through the bushes. Even though the name of the creek was Elk Horn Creek, the elk had been absent from the lowlands in recent years, preferring to stay away on the heavily forested ridges. The elk must be coming down to drink now; it was a good sign.

Sword ferns draped over still, deep pools. Brown spots of spores lined up on the underside of each frond, each precisely in line, each perfectly spaced, each the same exact size, hundreds on each leaf, millions along the bank of the creek. Was there reason to such precision?

In a shallow reach farther downstream, the gurgle of water caught my attention. Silver splashes erupted from the black water. Must be a submerged branch, I thought. As I approached, I noticed a conspicuous slapping motion of the water. Two splashes nearby seemed to be shifting location. Couldn't be submerged snags. Then I saw clearly the dark forms of pairs of submerged fish, thirty inches long. Coho salmon spawning! With powerful thrusts of their bodies, they were carving out depressions in the gravel, a male and a female working over each redd. These salmon, after being born in this same gravel bed, had spent three years in the open Pacific Ocean. In recent years coho salmon had been listed as endangered. Like other open ocean fisheries in the north Pacific and Atlantic, the coho salmon had been greatly reduced in number by logging, farming, and overfishing. I had not noticed a coho in this stream for almost ten years.

Now, in a display of incredible navigational skill, partly based on following the earth's magnetic field and partly based on detection of a unique chemical fingerprint from each stream, the coho had returned to the gravel bed of their birth. I silently watched the churning waters.

The realization that our brains simulate reality from a hidden field of information common to all individuals was the eureka moment. The deep reality of the universe is a hidden field of information that perforates time and space. Individuals with similar brains believe they are experiencing the same reality—the same universe—because it is sculpted from the same field of information.

The hidden field of information explains how each of us, while participating in the creation of a personal reality, shares common experiences with others, indeed, explains how we experience each other.

We participate in the formation of our own personal reality in a fashion analogous to our participation in the formation of rainbows.

Suppose a group of individuals, all in their own cars, has stopped on the side of the road to view a stunning rainbow. Each individual believes they all see the same rainbow, the same size arc, the same shades of color, and the same variations of brightness. Any given individual can point to a particular feature of the rainbow, and all others immediately recognize it. But when each individual gets back into his or her car and drives off in different directions at different speeds, a rainbow races along beside each car always the same distance away, the same speed as the car. Each individual experiences his own individual rainbow.

Likewise, each of us exists in our own vehicle of space-time, enveloped in a shell of light of uniform velocity. Even as we accelerate to high speeds, light has the same velocity in all directions, the same in front of us as behind. Just as we can never reach the end of a rainbow, we can never go faster than the speed of light.

Because we live together on earth under relatively constrained conditions of speed, energy, and mass, each of our interpretations of reality is similar. Now, suppose our rainbow observers get into rockets instead of cars and accelerate off in different directions at speeds approaching the speed of light. Suppose further that each rocket has a surveillance camera that instantaneously broadcasts images from cabin to cabin. This cannot be done, of course, but just suppose it could be done for the sake of a thought experiment. The images of the clocks in the cabins of all spacecraft would appear to go slower and the occupants would age more slowly than any observer watching the images on a monitor in his own cabin. Time, mass, and energy are not the same for each of us. We each exist in our own reference frame of time, mass, and energy; that only becomes obvious at speeds approaching the speed of light. At slow speeds, we think we are all in the same reference frame, just as the group of individuals that stopped their cars by the side of the road think they all see the same rainbow.

The rainbow is created when sunlight, always directly behind the observer, is reflected and refracted in water drops, always directly in front of the observer—just as light is reflected and refracted inside lead crystals hanging in a window. The angle between the sun, the rainbow, and the observer is always forty-two degrees. If the observer is stationary, the rainbow is stationary, though the water drops are moving. The rainbow follows precise laws, a hidden field of information.

A bluebird sitting on your shoulder sees the same rainbow as you. But if the bluebird flies off, its rainbow leaves and yours stays. As the bluebird circles, its rainbow appears to fly over you. The bluebird may wonder why its rainbow flies over you. Moral of the story: Don't envy the bluebird.

Is the rainbow an illusion, or is it real? The rainbow exists for anyone looking into a rainstorm with the sun shining directly behind them. It is a predictable outcome of precise conditions. Within the context of those conditions, the rainbow is a reality. The reality of the rainbow is contextual.

Is our simulation of reality an illusion? It is an outcome of experiencing the hidden field of information with a particular brain filter. Our personal reality is simulated from a fraction of the total field of information. It is a predictable outcome of precise conditions. Within the context of a given brain function, our personal reality is not an illusion, but it is contextual. There is a deeper, grander reality, the hidden field of information.

This deep reality is independent of the observer. However, the observer participates in the creation of a personal reality by experiencing discrete aspects of the hidden field of information. Some choices are hardwired into the physiology of the brain; some are imposed by experience.

Our conscious worldview is like the desktop on a personal computer screen. The programs and stored information of the computer are like our unconscious minds. The screen is user-friendly, with organized files and programs we can open with the click of a mouse. The desktop does not reveal the huge amount of information processing that goes on in the hard drive for each click of the mouse. Our consciousness is not aware of the automatic functions that are carried out in the unconscious mind and the multitude of information that is filtered out before we are presented with a conscious perception of the world.

The field of information is the same for all of us, but each of us interprets the information differently. We cannot load the dice, but we can decide which game to play; we cannot change the field of information, but we have freedom in the way we make use of the information.

Plato, the Greek philosopher, proposed that there is a hidden world of abstract forms that guide geometric and physical laws. The universal law of gravity discovered by Newton three centuries ago is one aspect of the hidden field of information. Niels Bohr showed that each electron orbiting the nucleus of an atom follows hidden quantum waves. The hidden field of information, in the form of quantum waves, prevents electrons from collapsing into the nucleus of atoms. Each of the incalculable number of atoms in the universe subscribes to the same information. If quantum waves are the foundation of the atom and dictate the behavior of electrons, why could they not be the field of information that underpins the universe?

The hidden field of information pervades nature and guides living forms as well as inanimate objects. Hundreds of single-celled animals in every drop of water—amoebas, parameciums, and dinoflagellates—navigate toward light and food with remarkable coordination of cilia and flagella. What information are they using to coordinate their movements? A soldier with his leg cut off feels the missing appendage even though it is gone. Lizards, crabs, and salamanders can regrow legs. Are they following the same energy blueprint felt by the amputee war victim?

We are often told that DNA carries all the information for growth and maintenance of living things. Yet DNA cannot guide which cells will become bone, which become skin, or which become muscle as a salamander grows a new limb. We are told that chemicals are the guides. But how do chemicals align themselves in the proper configurations such that skin will form on the outside of bone, for example.

The biologist Rupert Sheldrake has proposed that as-yet-undetected morphogenetic fields act as templates for the formation of missing appendages and guide the formation of the six million molecules of DNA or the billion trillion proteins produced in our bodies every second. The production of these complex molecules is the most precise process we know of. Is Sheldrake's morphogenetic field another component of the hidden field of information?

At the turn of the nineteenth century, scientists thought that the universe was made of matter. In the middle of the twentieth century, the universe was thought to be made of energy. Now, at the turn of the new millennium, it is becoming evident that the universe is made of information.

How is it possible for the foundation of the universe to be composed of nothing but information? How does information make something out of nothing?

The field of hidden information is like the invisible radio waves that guide ships at sea. Suppose we are flying at high altitude over a calm sea with several dozen ships dispersed, undetectable, beyond the resolution of our perceptions. Suppose each ship receives a radio message to navigate to a common location. Once aggregated, the ships form a mass large enough to be detected, a mass that seemingly appeared out of nowhere. If we were unaware of the radio signals, it would appear that a force drew the ships together. The quantum potential is a field of information that directs the position of subatomic particles just as radio navigation beacons direct the location of ships at sea.

Matter, manifesting from the zero-point energy, is a mostly empty aggregate of infinitesimally small mass-less particles that follow rules of motion, position, and symmetry conveyed by the field of information. At the smallest scale, like sea foam adrift in the bow wash of the ocean liners of subatomic particles, the eternal interconnected system of black holes, white holes, and wormholes makes up the fabric of the universe beyond space-time.

The hidden field of information that is the basis of reality is a plausible means of connecting the world with our minds. Connections in the field of information are made when aspects of quantum waves come into resonance, vibrating in phase, phaselocked. Some quantum waves are entangled out of phase, destructively interfering such that they blink out of existence. Through a system of phaselocked quantum waves, the field of information behaves as a single entity. If you change a wave here, it instantaneously interacts with all other waves in the field of information, just as moving one side of a seesaw immediately moves the other end. Perhaps quantum waves are held in phaselock by the zero-point energy in the quantum vacuum just as pieces of fruit are held in place by the gelatin in a Jell-O salad. When we shake the bowl of Jell-O, waves of motion cause all pieces of fruit to move instantaneously. Since the field of information is interconnected, our minds have access to all parts of it.

Is the field of information the objective deep reality that Einstein sought? According to the theory of relativity, matter cannot travel faster than the speed of light. The collapse of the wave function must travel faster than light; therefore, quantum waves cannot be matter. As demonstrated by Niels Bohr, they must be waves of probability. Bohr's God plays dice. But is information matter? In the brain-filtering process, the wave function does not collapse. One aspect of the wave function is experienced, but the other aspects continue to exist in the hidden field of information.

If basic reality is interconnected, why do we feel isolated and alone? How do we experience local events in an interconnected, nonlocal universe? Once again, it's a matter of choice. When nonlocal aspects are filtered out, we experience the complementary function locally. For example, in the rainbow analogy, the density or size of raindrops could be altered nonlocally, from a distance, without altering the reality of the rainbow to a local observer.

• • •

If time is an illusion, why does it seem so real? One possible explanation is that the fundamental field of information is beyond time. Our interaction with the field of information creates time by filtering out its complementary aspects. Perhaps the timeless field of information is sprinkled with time capsules such that when we experience one, time starts up.

In his book *The End of Time*, Julian Barbour proposes that there is only one universe and it consists of a stack of moments that, like a deck of cards, can be shuffled and reshuffled arbitrarily to give the illusion of time. These many instants of time are analogous to files of real-time movies on the quantum-field computer desktop. We click a mouse on a file, and the movie starts. The static hard drive is the hidden field of information. The desktop corresponds to the simulation of consciousness. When we engage a cosmic time capsule, it opens up a new space-time event that engages our consciousness to the degree that it doesn't pay attention to the unconscious part of us that opened the time capsule.

Our minds are capable of interpreting a series of closely related images as continuous motion. Electrons that sweep across a television screen or computer monitor illuminate one spot at a time and make a full sweep of up to ten thousand dots in less than a thirtieth of a second. Our mind interprets the minuscule changes in position of objects on a sequence of static screens as smooth change. The motion of elementary particles is really a series of particles that pop out of the quantum vacuum one after the other. Our minds interpret the sequence of many individual particles as the motion of a single particle. Similarly, our minds create reality by interpreting many static instances as a continuous reality.

A few unfortunate individuals suffer from a type of brain damage that prohibits them from seeing continuous motion. Rather than seeing a cup of coffee being continuously filled, they see only static images—now empty, now full.

When we prick our finger with a pin, the act is not registered in our consciousness for up to half a second after it occurs. We unconsciously yank the finger away before we say, "Ouch, that hurts." Yet the mind, like a talk show host who bleeps out obscene remarks with a delay button, gives us the impression that the experience of an event takes place at the same time as the event. We become aware of the pain at the same time that we become aware that the pin has pricked us, even though we have already yanked the finger away. The mind can fool us into believing that two events separated by half a second occur simultaneously. This is known as the backward referral of subjective experience. Might the mind be capable of making us believe that time exists when it really does not—of fooling us into believing we see motion when there is only a series of closely related static moments?

Perhaps deep reality contains a series of many instances that pop up out of a timeless field of background information. Our minds assemble a selection of instances. We become conscious of a continuous replay that instills the sense of time and space in us. We experience one aspect of the wave function. The other aspects do not spin off into universes but remain embedded in the field of information, in different instants of time that we do not experience. Even though the field of information is timeless, time is not an illusion within the context of our experience of the field of information.

Julian Barbour proposes that there are rules that, to some extent, govern the sequence of instants that we can experience. Similar instants are more likely to be experienced one after the other than wildly disparate instants. The instants we are most likely to experience in sequence are those most likely to produce the illusion of continuous motion and time. Once we discover the rules, we can exert conscious influence over which futures we experience.

The hidden field of information gives the most exciting explanation of the near death experience. Those aspects of the mind that normally track a course through the sea of information are abandoned near death. The detached experience allows the nonmaterial mind to experience the time-free and interconnected aspects of the field of information. Similar mind states can be achieved in dreams, under the influence of psychedelics, during hypnosis, and in other trancelike states.

Why don't we experience near-death reality all the time? The May 29 experiences convinced me that the reality experienced by brain-generated consciousness is only a pale representation of what goes on in the universe. Similarly, our senses tell us that the earth is flat because they cannot grasp the full extent of the earth's girth. However, traveling around the earth in a jet, we can generate a series of images and piece them together into a map that shows the true nature of the earth. Some planets in the night appear to move backward some of the time. Using mathematical equations of orbital motion, Johannes Kepler demonstrated that planets appear to move backward because the earth moves forward, overtakes a planet, and produces the illusion of backward motion.

The brain's preoccupation with maintaining the automatic and emotional functions of life requires it to filter out information from most of the universe by interpreting it in a way to best sustain local biological survival. Fortunately, we are able to direct our focus of attention, scan the many aspects of reality, record small portions at a time, and combine the pieces into a total perception of reality. At the time of death, normal brain functions are abandoned, and we experience an expanded version of reality, enabling an expanded experience of the field of information—the many aspects of the wave function accessed by the nonmaterial mind.

How exactly human consciousness focuses on one aspect of the wave function is the most outstanding problem in science and human cognition.

The universe is made of infinite possibilities inscribed in wave functions at all levels of matter. In the bountiful many-instants universe, every little "could be," no matter how improbable, gets its time to shine. However, the very act of observing limits us to observing what we are trying to observe. At various levels of mind, we make choices as to what to experience, and that process of choice creates time so that everything doesn't appear to be happening at once.

Letting go of the brain function is the key to experiencing expanded awareness, but we cannot "try" to let go. When we "try," the ego is in control; the nonmaterial mind focuses through the brain. Likewise when we "try" to measure scientifically the direct presence of the nonmaterial mind, it collapses and changes, as happens when we open a box with Schrödinger's cat inside.

It's like throwing crumpled-up wads of paper in the trash while sitting at a desk. At first we focus on the basket; some wads go in, some don't. We become absorbed in the act of writing, forget to look at the wastebasket, and don't listen to that part of the brain that says it's impossible to hit a wastepaper basket without our looking at it. Amazingly, wads then go in without looking. We're in the wastebasket disposal zone. Things like that happen all the time. As soon as we say, "Wow, did you see that?" and call a friend or lover—"Hey watch this!"—of course, it doesn't go in. The wave function collapses; possibilities decrease; the magic disappears. If we want it to happen, it won't happen. But it's not a superstition. It's a state of quantum reality. Wanting something limits the wave function by focusing on only one aspect, and like King Midas, our own touch isolates us from the world of expanded quantum waves in the timeless field of information.

Everyday consciousness is a hypothesis, a working model, an outgrowth of natural selection that has allowed us to adapt to our physical environment. It has served us well, but who knows what adaptations will best serve our future evolutionary needs? Perhaps we will learn in the future to develop a more extensive communication between our conscious and unconscious, our material and nonmaterial minds.

I am not the first person to realize that the mind survives the body, or that the reality of the universe is a marvelous field of information and infinite potentials, or that we ourselves create time by opening static time capsules in the field of information. But I had the joy of discovering these ideas independently before I was exposed to them by others. The meaning of my May 29 events is clear. I finally understand, and the process of understanding has been exhilarating. As a route tracker, I have tried to piece together a preliminary map of the deep reality of the universe, an expanded view, one full of adventure, promise, and hope.

The ferry slipped into the waters of Long Island Sound. I sat in the upper deck scanning the distant shoreline of Martha's Vineyard, trying to spot the house Woody had built on the shoreline. I wondered what it would be like to see him again. Would he be happy to see me, and what about Norman? They were both dedicated patriots who had served their country in World War II; I had demonstrated against the war in Vietnam after we had been together on Mount Everest. I was no longer the conservative Yale graduate student they had known forty years ago. Woody's daughter, Jen Sayre, had warned me that sometimes Woody, now eighty-two, was cantankerous.

The ferry pulled into the slip. As soon as I set foot to pavement, I recognized Hans Peter and gave him a hug.

He was well tanned, trim, with a full head of hair even at age sixty-three. He had been sunning himself on Woody's beach for over ten days, having flown in from Switzerland. He doesn't get to the ocean often and was enjoying Woody's beach access.

A lanky woman with short-cropped hair, Woody's second wife, introduced herself, "Hello, Roger, I'm Pat Sayre, welcome to the Vineyard." She opened the rear doors of her carryall, and I threw in my luggage.

"How's your wife, Maggi, and your boys?"

"They're all doing fine, back in Oregon." I climbed into the passenger seat, and Pat slid into the driver's seat.

Hans Peter leaned over from the backseat. "How was your trip? I heard you almost didn't make it. We're amazed that you did."

"I had to wait in line for five hours at PDX. By the time I got to the ticket window, I had missed my flight to Minneapolis. I waited another two hours for a flight to Detroit, then two hours for a flight to Providence. By the time I got to Woods Hole last night, I had missed the last ferry. But this was trivial compared to what others went through. What time does the reunion start?"

"It started half an hour ago, but everybody's waiting for you," replied Hans Peter.

I was surprised when we pulled up to the Martha's Vineyard Senior Center. "I thought we were going to meet at Woody's house."

"No way," said Pat. "Wait until you see all the people here."

Jen Sayre met me at the entrance, threw an official "Four Against Everest" T-shirt at me, and led me to the men's room where I changed.

When I emerged into the meeting room, all the seats were taken and a few people were standing along the walls. I spotted Norman. We greeted each other warmly.

Woody was hunched over in a wheelchair in front of a large projection screen at the front of the room. I approached him and bent over to shake his hand. "Hello, Woody. How are you?"

He looked up at me, eyebrow raised painfully high, as if it were a huge effort to remain conscious. His blue eyes were clear through his glasses, and he said unintelligibly "Oh, I didn't——— with the beard." He shook my hand with gnarled, stiff fingers.

As my mind reconstructed what he'd said, it stopped its replay on "beard."

Immediately I thought he was criticizing my beard. My heart

went cold toward him. I said to myself, *I hope he doesn't accuse me of being a fuckin' beatnik.*

"Oh . . . ah . . . Okay. Good to see you." I straightened up, turned on my heel, and replayed what he had said several times in my mind. Beard, beard, something about a beard. Oh, that's it; he said, "He didn't *recognize* me with the beard." Oh, shit. I turned back to him, but he had relapsed into a minor stupor, his head down on his chest.

Norman, Hans Peter, and I stood behind Woody while Jen introduced us, then we took our seats near the projectors. Woody rolled a few feet out onto the floor, raised his eyebrows, and said in a quivering voice, "Well, who would have thought we would have had such a crazy idea as to try to climb Mount Everest." He rolled back and turned to Norman, who started the projector. We sat through a series of slides and a short sixteen-millimeter movie, which interested me most. Woody had taken movies of the terraces he had seen. They were real and unexplainable, flat platforms of dark rock bordered along the front edge by bands of white. I had never seen such rock formations, but they are probably the effect of weathering somehow turning black rock white along its exposed edge.

I got up and publicly thanked Woody for making the adventure of a lifetime possible. The audience applauded. Woody lifted his arms high over his head and let out a wild whoop, "All Right!"

Randy McNeely, a climber and filmmaker who I later discovered was making plans to retrace the route of our expedition and make a film of it, stood up and said, "The historians of Everest claim that the Sayre Expedition was extremely lucky with the monsoon. If it had arrived a few days earlier in the spring of 1962, they wouldn't have made it. But let's face it. Some people make their own luck."

They don't know the half of it.

By a strange coincidence it was my birthday September 16, 2001, just five days after the tragedy of the attack on New York's World Trade Center and the Pentagon. Jen brought out a birthday cake shaped like Mount Everest, and the whole group sang happy birthday to me. Tears came to my eyes. After all we had been through, the hassles with the State Department over passports, the blackballing of our expedition by the members of the 1963 expedition and the

American Alpine Club, this turnout! This recognition! I was choked up with gratitude.

Afterward we signed copies of *Four Against Everest* and the reunion T-shirt. I felt like a rock star, bathing in the warm glow of public recognition for the first time. Then it was time for group photos. I tracked Hans Peter down at the food table; he was stuffing shrimp into his mouth. "I never had these before; we don't have these in Switzerland!"

"Wait until you have a lobster!" I said.

After the reunion, Hans Peter and Jen showed me the room at Woody's house where I would stay—Woody's old office lined with philosophy books.

"Let's go to the beach," said Hans Peter.

"Not me; I'll see you guys later," said Jen. She worked as a massage therapist and had to attend to one of her many Vineyard clients.

I put on swim trunks and grabbed a towel.

Hans Peter led me through Woody's house, a spacious A-frame with two bedrooms. Woody sat in his favorite armchair near glass doors that led out to the deck and the open sea. His head tilted back, mouth wide open.

"Woody," Hans Peter addressed him in a loud voice, "Roger and I are going out to the beach."

I stood back, half hoping Woody wouldn't notice me.

Woody closed his mouth, raised his eyebrow in recognition, lifted his crippled hand, and murmured a barely audible, "Okay."

We followed the narrow path through beach grass. The day was still, warm, and the sun shown brightly.

Hans Peter and I lay on the narrow beach, which was covered with sparse dune grass, cobbles, and algae.

"Well, Roger, here we are again," he said.

"Yes, it's a miracle—almost like we'd never been apart. You look really fit."

"I'm still guiding groups in the Alps to stay in shape, going barefoot to connect to the land." He lay back on his towel, closed his eyes, and tilted his head to the sun.

A seagull appeared, winging along the shore with a clam in its beak. Right next to us it flapped, hectically ascended straight up to a significant altitude, and then dropped the clam on the line of boulders along the water's edge. The seagull dropped down, picked up the clam, and dropped it again.

"Look, Hans Peter, the gull is trying to break open the clam," I said.

"Yes. I know. It's here every day."

"So I heard you were in Bolivia in 1968."

"Yes, I've spent a lot of time there, as part of the Swiss development program. And while I was there, I climbed Huascarán, the highest mountain in Peru."

"What a coincidence. I climbed it too." Paul and I had had a wonderful climb of Huascarán, financed out of our own pockets for forty dollars.

"I'd like to return to Peru and study the glaciers there," I said. "I think I found evidence that some of the Inca ruins predate a period of glaciation."

"Wonderful; let's go together."

"All right!"

The seagull dropped the clam and it cracked open. Immediately dozens of other seagulls converged, squawking, flapping their wings, and surrounding the clam dropper as it frantically pulled a juicy piece of clam meat out of the shell and flew off. The other seagulls surrounded the empty clam shell and fought over the remains.

That evening the four of us were reunited at Woody's house. We sat around a table and watched the fire in the fireplace. Jen sat with us.

"What's Woody been doing with himself?" I asked.

"After he retired, he became obsessed with blackjack, attended tournaments in Vegas, all over the place." Jen placed a half dozen T-shirts on the table for us to sign. Norman, Hans Peter, and I set to work, but Woody stared down, ignoring the pen Jen had put in front of him.

"He's being stubborn," said Jen.

"Here, give me that thing." I snatched Woody's pen. "I'll forge his signature."

Woody looked up, smiling, and reached for the pen, his hands trembling, and began signing the T-shirts. He relapsed into another state of consciousness and stared at the fire, apparently not listening to our conversation.

"He's really happy you're here. He asked about you every night for the past two weeks," said Jen.

"Well, I'm happy to be here. You know *Four Against Everest* is very popular on the Internet. Several sites have a quote from the book."

"What quote is that?" asked Jen.

"I don't remember it exactly; something about one breath of air is not enough."

Woody, head still down, murmured, "One can't take a breath large enough to last a lifetime; one can't eat a meal big enough so that one never needs to eat again. Similarly, I don't think any climb can make you content never to climb again." Woody looked up at me, raised his eyebrow. He gave me a thumbs-up. The fire danced in his eyes. He turned to Jen and said, "I'm ready for bed now. Good night, everybody." Jen wheeled him off to the bedroom.

Woody led me on my first great adventure, but it was no greater than the adventure of love and childbirth led by Maggi, the adventure of social awareness led by Paul, or the adventure of scientific discovery led by my father and brother, Stan. Now it is clear to me that the greatest adventure lies at home in each of us.

If my interpretation of the May 29 experiences is at least in part correct, then in the future we will have a taxonomy of mind states, an understanding of the levels of consciousness, and a means of exploring the niches of the unconscious. We will fuse the wisdom of shamans with the techniques of science to construct a blueprint of consciousness.

Simultaneously, we will map the terrain of the hidden field of information, the branching valleys, the flow of probabilities, the quiet backwaters of slow time, and the endless bounty of possible futures. Using earth as our training ground for the adventure of inner space, we may be able to identify the phaselock codes that generate demons and those that preordain angels. We will discover pathways from place to place, and trace the network of wormholes that burrow beyond time and beneath space.

Most important, we will eventually discover how mind interacts with the field of information, how consciousness collapses the wave function, or how we create the flow of time by opening time capsules.

Once we have that information, we will finally have the knowledge and freedom to track our own realities. We will be able to experience any future from an infinity of possibilities by bringing it into consciousness. With love and caring, we can help each other deal with the awesome possibilities. Like the navigators in *Dune,* we will be able to experience any place or any time at home. Our dreams will become reality. We will be able to direct the flash-light of consciousness to wherever we want. Now, *that* will be a great adventure.

ACKNOWLEDGMENTS

In a work that combines personal narrative with quantum mechanics, the physics of time, and consciousness research, it is impossible to acknowledge all who deserve it. Nonetheless, I wish to highlight the contributions of a few individuals who have greatly influenced me. The words and works of Erwin Schrödinger have been an inspiration to me for more than forty years. His insights are as meaningful today as they were when he presented them in the first half of the twentieth century. His concept of quantum entanglement led directly to the idea of the phaselock code presented in this book. The many-worlds interpretation (MWI) of quantum mechanics developed by Hugh Everett and popularized by Bryce DeWitt, John Archibald Wheeler, and David Deutsch led to the idea that we have some degree of freedom in choosing our futures from an infinity of possibilities. Julian Barbour's book *The End of Time* provided the perfect explanation of my experiences through his concept of the many instances of time. The amazingly prescient work on consciousness by William James led to the idea that our personal realities are sculpted from a field of information. This idea was supported by the research of Benjamin Libet and popularized by Tor Nørretranders in *The User Illusion* and David Chalmers in *The Conscious Mind*.

I wish to thank the family and friends that shared these adventures with me. Any character sketch is incomplete and, out of necessity, I had to limit my descriptions to my personal impressions recovered from limited memories in a specific setting over a brief period of time. I regret that I had to limit the full breadth and depth of any character portrait to a few words in a specific narrative line; this does not mean that I hold such a limited view in general or wish to imply that the limited view represents the full range of any personality's character and experience.

I wish also to acknowledge those who helped me learn the craft of the written word: Fred Pfiel, Ken Babbs, and Ken Kesey along with those who read and critiqued early drafts of the manuscript, Richard Kennedy, Ed Cameron, Tracy Smith, Dennis Curran, Susan Pilling, Vaughn Marlowe, and Elizabeth Rose.